Production Parameters in Textile Industry

Production Parameters in Textile Industry

Dr. Sanjay Sharma

CRC Press
Taylor & Francis Group
Boca Raton London New York

CRC Press is an imprint of the
Taylor & Francis Group, an **informa** business

WOODHEAD PUBLISHING INDIA PVT LTD

New Delhi

First published 2025
by CRC Press
4 Park Square, Milton Park, Abingdon, Oxon, OX14 4RN

and by CRC Press
2385 NW Executive Center Drive, Suite 320, Boca Raton FL 33431

CRC Press is an imprint of Informa UK Limited

British Library Cataloguing-in-Publication Data
A catalogue record for this book is available from the British Library

Print edition not for sale in South Asia (India, Sri Lanka, Nepal, Bangladesh, Pakistan or Bhutan).

ISBN13: 9781032954967 (hbk)
ISBN13: 9781032954974 (pbk)
ISBN13: 9781003585183 (ebk)

DOI: 10.1201/9781003585183

Typeset in Times New Roman
by Bhumi Graphics, New Delhi

WPI

Contents

Preface

In the context of production or manufacturing, the related engineering/management lies in producing certain items. Efficiency might be at a stake if a precise focus is not on the total cost or associated production lot size. An entrepreneur or entrepreneurial engineer/manager, with special reference to production or manufacture, frequently enters into an arena where the related parameters might often fluctuate. Under such scenario, it is also necessary to understand various production parameters and their influence as well as their interaction. The present book on "Production Parameters" explains such aspects in detail along with the rigorous analysis and several examples.

Additionally, some of the salient features are concerning the fresh approach and uniqueness of this book. All the production parameters, their estimation and effects are covered for the first time. Similarly, their interaction has been especially analysed in detail. It is expected to be highly useful for engineers/professionals/M.Tech./MBA students interested in the production/manufacturing scenario.

Sanjay Sharma
Mumbai, India

1
Introduction

Abstract : This chapter presents an introduction and understanding of various production parameters, which is necessary for knowing the implications pertaining to their variation among other issues. Such parameters include a customer demand and also the costs of facility setup, inventory holding as well as production. A cycle time, production rate and shortages are also included among the relevant production parameters. After an explanation for these factors, the total annual cost is derived for obtaining an economic batch quantity. It is also extended to include shortages. An interaction of production parameters is discussed with the help of relevant tables.

Keyword : Production parameters, batch quantity, total annual cost, interaction of parameters

In any factory, raw material is being converted to the finished product with the use of certain facility. This facility represents the production process. In the context of production management, certain parameters play very important role. A detailed understanding of these production parameters is necessary for knowing the implications pertaining to their variation among other issues.

1.1 Production parameters

Different types of production parameters are shown in Fig. 1.1.

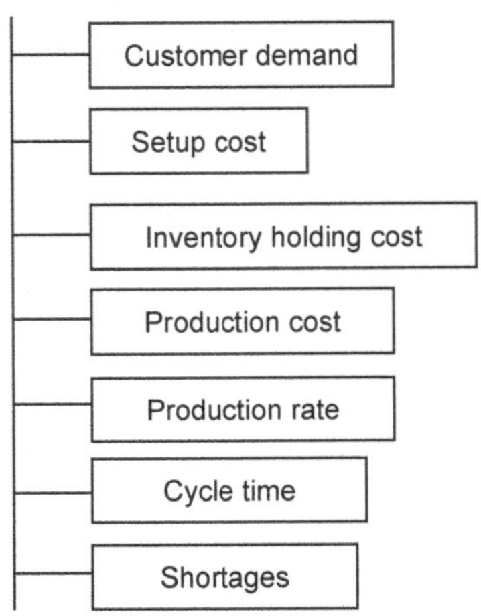

Figure 1.1: Types of production parameters

1.1.1 Customer demand

Depending on the end customer demands, the retailers/dealers place the orders for certain quantities of a product to the manufacturing company. For instance, the demands of a manufacturer or producer are shown in Table 1.1. Each month, there is a constant demand of 50 products.

Table 1.1: Uniform demands.

Month	1	2	3	4	5	6	7	8	9	10	11	12
Number of products	50	50	50	50	50	50	50	50	50	50	50	50

In this example, annual demand is:

$$12 \times 50 = 600 \text{ products}$$

Fig. 1.2 shows the graphical representation of uniform demand.

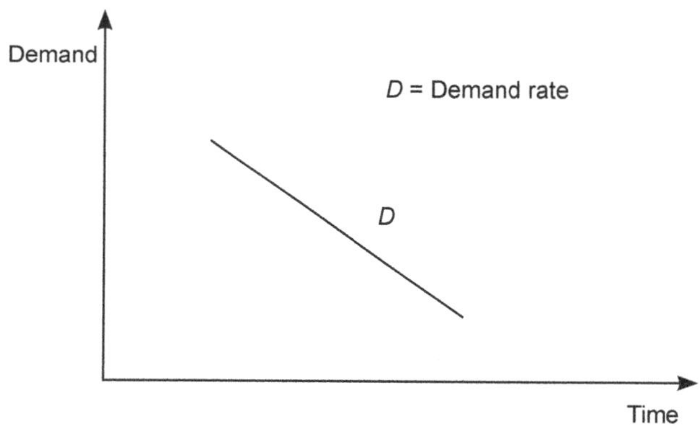

Figure1.2: Representation of uniform demand

There are various alternatives to fulfill the annual demand. Such options might be analysed on the basis of annual total cost. A feasible/practical outcome may be implemented which is also expected to probably yield the least total annual cost.

1.1.2 Setup cost

The following activities might be associated with the setup of any facility in a production company:

(a) Work piece, on which certain machining work is to be done, will be held in certain job holder. Once this is set, it is useful for similar such work pieces one after another.

(b) There are various types of cutting tools and such cutting tool is fixed in certain tool holder. Once this is set, it is useful for producing a number of similar items.

(c) Facility/machine parameters are to be set, e.g., speed. Production speed is also dependent on such parameters among other factors.

Efforts associated with the setup of facility should be listed and the costs corresponding to those will give an estimate for setup cost.

1.1.3 Inventory holding cost

A production facility manufactures certain product at a specified rate of manufacture. If the demand rate is less than the production rate, then there is excess number of products available at the end of period. This may be referred to as the inventory. When we hold the inventory for certain time, there are inventory holding costs. Storage space will need to be arranged for keeping the produced item inventory. In case of the need, items should be preserved also. Production cost is incurred which is unused until the item is sold. Such related aspects should be listed and inventory holding cost per unit product should be estimated for certain specified period.

1.1.4 Production cost

After purchase of raw material, the production facility will convert that to the finished item. While producing certain item, the following activities are relevant:

(a) After setup of the facility, run it for actual manufacture of the product.

(b) Human resources time for actual production

(c) Associated material handling

Depending on the situation, relevant efforts are converted to the cost in order to find out the production cost per unit item.

1.1.5 Production rate

Suppose that a facility runs at a certain speed, and it produces 60 units per month. The production rate can be considered as follows:

60 units per month, or $12 \times 60 = 720$ products per year.

In other words, annual production rate is 720.

Production rate may also be visualised as shown in Fig. 1.3.

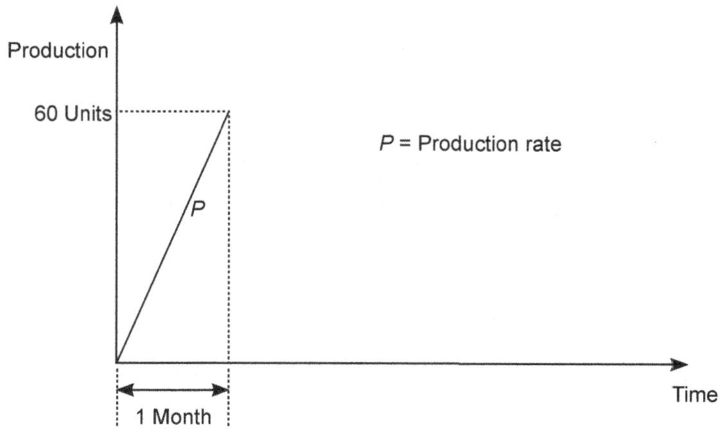

Figure1.3: Production rate

Suppose that there is no demand in a particular month, there would be a stock of 60 products at the end of month. However, as discussed before, if the uniform demand rate is 50 products per month, then stock will be of 10 products at the end of the month. This is because:

$$60 - 50 = 10 \text{ products}$$

In other words, stock or inventory build-up rate is 10 products per month. It is also shown in Fig. 1.4.

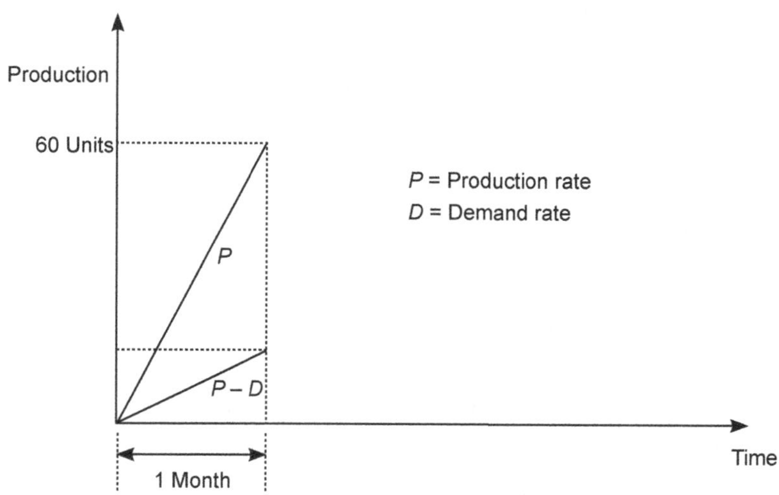

Figure 1.4: Production rate and stock build-up rate

1.1.6 Cycle time

Consider the example of constant demand of 50 products per month as shown below:

Month:	1	2	3	4	5	6	7	8	9	10	11	12
Number of products:	50	50	50	50	50	50	50	50	50	50	50	50

As the annual demand is 600 products, there is a need to produce at least 600 products in one year. Suppose that the manufacturing company decides to produce 200 products in one setup and the production rate for this product is 200 units per month, then there will be a need of the three facility setup in one year. This is shown in Fig. 1.5. Consumption or demand is available throughout the year. However, the production takes place only for one month in each cycle of 4 months duration. Therefore, the cycle time for this scenario is 4 months.

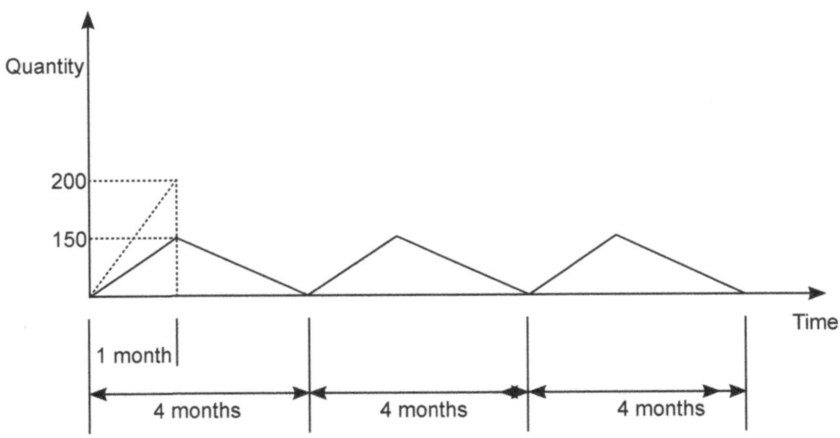

Figure 1.5: Manufacturing cycle

1.1.7 Shortages

As shown in Fig. 1.6, if production is not started as soon as stock becomes zero, then there are shortages. In other words, demand of the product exists but availability is not. These shortages may be backordered, i.e., the next production cycle may compensate for such shortage in quantities in addition to regular quantities. There are costs associated with the shortages and shortage cost per unit product should be estimated for a specified period.

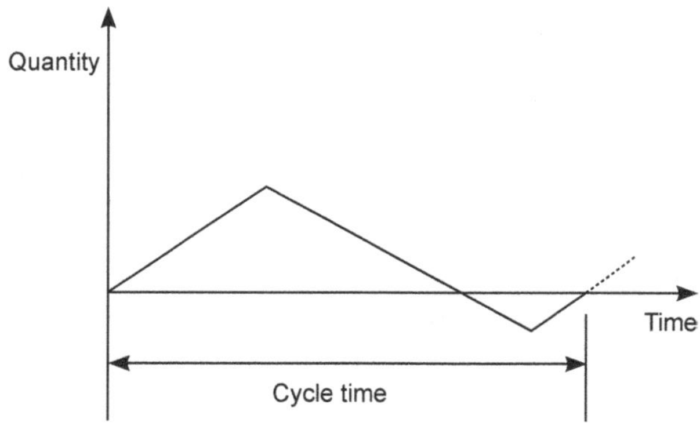

Figure 1.6: Representation of shortages

1.2 Economic batch quantity

As shown in Fig. 1.7, batch quantity or batch size Q is produced in a manufacturing or production time.

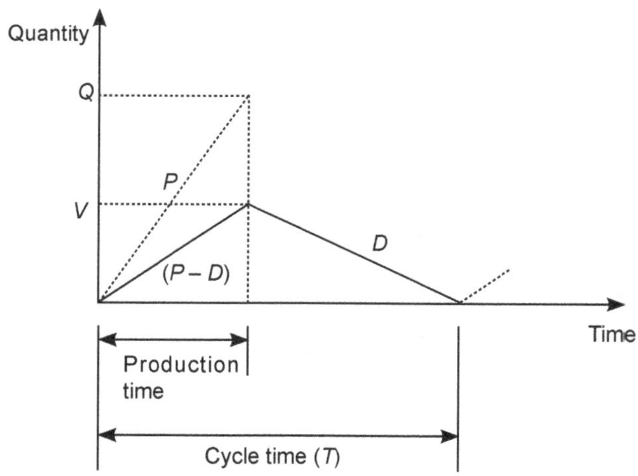

Figure 1.7: Production cycle

Let:

P = Annual production rate or production rate/year

D = Annual demand or demand rate/year

For example, if batch size Q is 200 products and production rate is 720 units per year, then

The production time $= \frac{200}{720} = 0.278$ year

In order to generalise, production time is equal to: $\frac{Q}{P}$

As the demand is available throughout the year and the production rate (P) is greater than the demand rate (D), inventory stock build-up rate is ($P - D$) during the production time.

Now,

V = Maximum stock during the cycle

And production time can also be expressed as: $\frac{V}{(P-D)}$

Therefore,

$$\frac{V}{(P-D)} = \frac{Q}{P}$$

Or
$$V = \frac{(P-D)Q}{P}$$

Or
$$V = Q(1 - D/P) \tag{1.1}$$

In a cycle time T, stock decreases after the production time is over and this cycle repeats.

As the stock increases from 0 to V, and then from V to 0, the average stock is $\frac{V}{2}$ and the annual inventory holding cost:

$$AIC = \frac{V}{2}.I \tag{1.2}$$

where I is the annual inventory carrying cost per unit

Substituting the value of V from Eq. (1.1),

$$AIC = \frac{Q(1-D/P)I}{2} \tag{1.3}$$

As batch quantity Q increases, annual inventory holding cost increases. With the annual demand D and batch production quantity Q in each setup or cycle, the number of setups in a year can be expressed as $\frac{D}{Q}$

Annual production setup cost is as follows:

$$APC = \frac{D}{Q}.C \tag{1.4}$$

where C is the fixed setup cost per setup.

As batch quantity Q increases, annual setup cost decreases.

In the present context, the total annual cost:

$$E = AIC + APC$$

Or

$$E = \frac{Q(1 - D/P)I}{2} + \frac{DC}{Q} \tag{1.5}$$

Total annual cost has the following two components:

(i) Annual inventory holding cost

(ii) Annual setup cost

Variation of Q has the opposing effect on these cost components. While decreasing the Q, AIC reduces. But ASC increases with a reduction in Q. An objective of the management is to reduce the total costs. Therefore, it is necessary to obtain the quantity Q corresponding to which the total cost is the least. In order to optimize the Eq. (1.5), differentiate with respect to Q and equate to zero.

$$\frac{(1 - D/P)I}{2} - \frac{DC}{Q^2} = 0$$

Or

$$\frac{DC}{Q^2} = \frac{(1 - D/P)I}{2}$$

Or

$$Q^2 = \frac{2DC}{(1 - D/P)I}$$

Therefore, the economic batch quantity:

$$Q^* = \sqrt{\frac{2DC}{(1 - D/P)I}} \tag{1.6}$$

Substituting the optimal value of Q from the above Eq. in the Eq. (1.5), the optimal total annual cost:

$$E^* = \frac{(1 - D/P)I}{2} . \sqrt{\frac{2DC}{(1 - D/P)I}} + DC . \sqrt{\frac{(1 - D/P)I}{2DC}}$$

Or

$$E^* = \sqrt{\frac{DCI(1 - D/P)}{2}} + \sqrt{\frac{DCI(1 - D/P)}{2}}$$

[It may be noted that the AIC and APC are equal at the optimal point.]

Or

$$E^* = \sqrt{2DCI(1 - D/P)} \tag{1.7}$$

1.2.1 With shortages

In the previous case, shortages are not included, i.e., the production activity starts as soon as the stock becomes zero. However, demands might be available, but the replenishment or the production may not start and the shortages are said to occur. This is shown in Fig. 1.8.

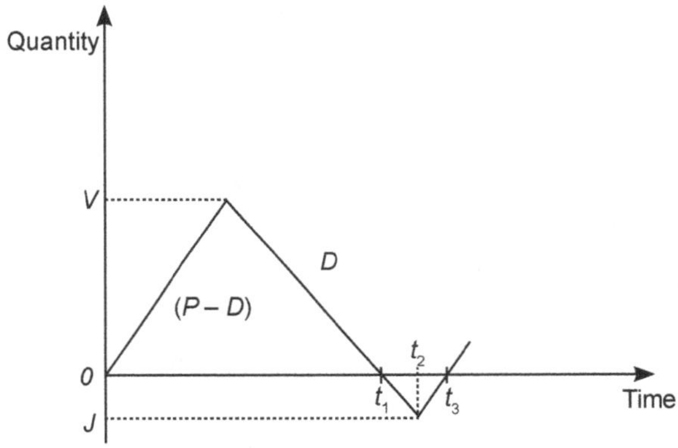

Figure 1.8: Inclusion of shortages

During the period t_1–t_2, demands are there, but the production activity is not yet started. At time t_2, the shortage quantity is maximum, i.e., J. During time $t_2 - t_3$, production activity corresponds to the replenishment of shortage quantities that are assumed to be completely backordered. Refer Fig. 1.8. Initially start may be from zero quantity. However, considering the continuous cycle/repetition afterwards, the case may be depicted by Fig. 1.9 also.

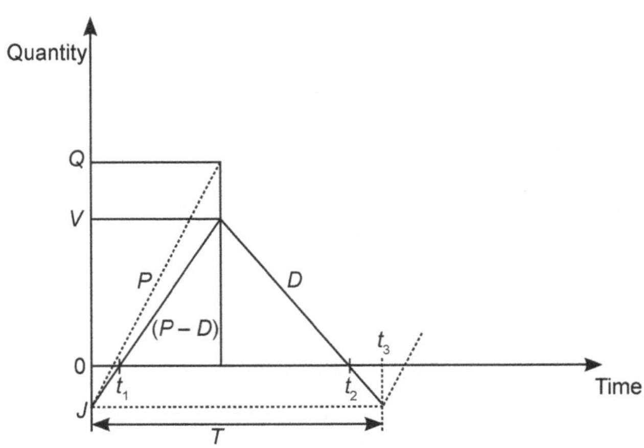

Figure 1.9: Production cycle with shortages

The shortages are during $0 - t_1$ and $t_2 - t_3$, in a production cycle time T.
Maximum shortage quantity = J
Annual shortage cost per unit = K

$$\text{Period during which shortages occur} = \frac{J}{(P-D)} + \frac{J}{D}$$

$$= \frac{J}{D(1-D/P)}$$

$$\text{Average shortage quantity} = \frac{J}{2}$$

$$\text{Number of production setup in a year} = \frac{D}{Q}$$

Therefore, the annual shortage cost is

$$ASC = \frac{J}{2}.\frac{J}{D(1-D/P)}.\frac{D}{Q}.K$$

Or

$$ASC = \frac{J^2K}{2Q(1-D/P)} \tag{1.8}$$

Positive inventory stock exists during $t_1 - t_2$, i.e.,

$$\frac{V}{(P-D)} + \frac{V}{D} = \frac{V}{D(1-D/P)}$$

Annual inventory carrying cost is as follows:

$$AIC = \frac{V}{2}.\frac{V}{D(1-D/P)}.\frac{D}{Q}.I$$

$$= \frac{V^2I}{2Q(1-D/P)} \tag{1.9}$$

$$\text{Now, production time} = \frac{Q}{P}$$

However, it is also equal to: $\frac{(V+J)}{(P-D)}$

Therefore, $$\frac{(V+J)}{(P-D)} = \frac{Q}{P}$$

Or $$V + J = Q(1 - D/P)$$

Or $$V = Q(1 - D/P) - J \tag{1.10}$$

Substituting the value of V in Eq.(1.9):

$$AIC = \frac{[Q(1-D/P)-J]^2 I}{2Q(1-D/P)}$$

Or $$AIC = \frac{\left[Q^2(1-D/P)^2 - 2QJ(1-D/P) + J^2\right]I}{2Q(1-D/P)}$$

Or $$AIC = \frac{IQ(1-D/P)}{2} - IJ + \frac{IJ^2}{2Q(1-D/P)} \qquad (1.11)$$

Annual production setup cost is

$$APC = \frac{D}{Q}.C \qquad (1.12)$$

In the present context, a total annual cost (E) is obtained by adding Eqs. (1.8), (1.11), and (1.12):

$$E = ASC + AIC + APC$$

Or $$E = \frac{J^2K}{2Q(1-D/P)} + \frac{IQ(1-D/P)}{2} - IJ + \frac{IJ^2}{2Q(1-D/P)} + \frac{DC}{Q}$$

Or $$E = \frac{J^2(K+I)}{2Q(1-D/P)} + \frac{IQ(1-D/P)}{2} - IJ + \frac{DC}{Q} \qquad (1.13)$$

For the least or optimal total cost, batch quantity Q and the maximum shortage quantity J need to be obtained.

Differentiating partially w.r.t. J and equating to zero:

$$\frac{2J(K+I)}{2Q(1-D/P)} - I = 0$$

Or $$\frac{J(K+I)}{Q(1-D/P)} = I$$

Or $$J = \frac{IQ(1-D/P)}{(K+I)} \qquad (1.14)$$

Substituting the value of J in Eq. (1.13)

$$E = \left[\frac{IQ(1-D/P)}{(K+I)}\right]^2 \left[\frac{(K+I)}{2Q(1-D/P)}\right] + \frac{IQ(1-D/P)}{2} - \frac{I^2Q(1-D/P)}{(K+I)} + \frac{DC}{Q}$$

Or $$E = \frac{I^2Q(1-D/P)}{2(K+I))} + \frac{IQ(1-D/P)}{2} - \frac{I^2Q(1-D/P)}{(K+I)} + \frac{DC}{Q}$$

Or $$E = \frac{DC}{Q} + \frac{IQ(1-D/P)}{2} - \frac{I^2Q(1-D/P)}{2(K+I)}$$

Or $$E = \frac{DC}{Q} + \frac{IQ(1-D/P)}{2}\left[1 - \frac{I}{(K+I)}\right]$$

Or $$E = \frac{DC}{Q} + \frac{KIQ(1-D/P)}{2(K+I)} \tag{1.15}$$

In order to obtain the optimal value of Q, differentiate w.r.t. Q and equate to zero.

$$\frac{KI(1-D/P)}{2(K+I)} - \frac{DC}{Q^2} = 0$$

Or $$Q^2 = \frac{2DC(K+I)}{KI(1-D/P)}$$

Or $$Q^* = \sqrt{\frac{2DC(K+I)}{KI(1-D/P)}} \tag{1.16}$$

Substituting the optimal value of Q in Eq. (1.14),

$$J^* = \frac{I(1-D/P)}{(K+I)}\sqrt{\frac{2DC(K+I)}{KI(1-D/P)}}$$

Or $$J^* = \sqrt{\frac{2DCI(1-D/P)}{K(K+I)}} \tag{1.17}$$

Substituting the Eq. (1.16) in Eq. (1.15), the optimal value of total annual cost can be obtained as follows:

$$E^* = DC\sqrt{\frac{KI(1-D/P)}{2DC(K+I)}} + \frac{KI(1-D/P)}{2(K+I)}\sqrt{\frac{2DC(K+I)}{KI(1-D/P)}}$$

Or $$E^* = \sqrt{\frac{DCKI(1-D/P)}{2(K+I)}} + \sqrt{\frac{DCKI(1-D/P)}{2(K+I)}}$$

Or $$E^* = \sqrt{\frac{2DCKI(1-D/P)}{(K+I)}} \tag{1.18}$$

1.3 Interaction of production parameters

Previous discussion refers to a derivation of economic batch size and total optimal cost. When shortages in the production system are not included, then the following parameters are relevant:

(a) Annual demand
(b) Production setup cost
(c) Production rate
(d) Inventory holding cost

1.3.1 Case without shortages

With an increase in demand, lot size tends to increase. However, if an optimal production lot size increases, then the operational settings such as storage space (including storage space between two facilities) might be insufficient. It may also lead to inconvenience in material handling. This is because material handling equipment and associated resources are compatible with the previous lot size requirement, and an adjustment with the revised lot size might not be that economic. There are situations in which similar lot size is preferred. Despite the most preferred demand increase, an objective of similar lot size can be achieved by:

(i) Reduction in the facility setup cost
(ii) Increase in the production rate

Depending on the ease in implementation, a suitable option may be chosen. Else an appropriate combination of more than one parameter variation might also be chosen. Similarly when variation is triggered by any parameter, a guide to management response with an objective of similar lot size is useful in practice as shown in Table 1.2.

Table 1.2: Response with an objective of similar lot size.

Variation triggered by	*Management response*
Demand increase	Reduction in setup cost/ Production rate increase
Demand decrease	Reduction in holding cost/ Production rate decrease
Setup cost decrease	Demand increase/ Holding cost reduction/ Production rate decrease
Setup cost increase	Production rate increase

Contd...

Contd...

Variation triggered by	*Management response*
Production rate increase	Demand increase/ Holding cost reduction
Production rate decrease	Setup cost reduction
Holding cost increase	Demand increase/ Production rate reduction
Holding cost decrease	Setup cost reduction/ Production rate increase

Similarly, with reference to a parameter change, total annual cost also changes. Increase in total cost might be a concern. Total cost also increases with the –increase in demand. However, this is preferred because it leads to the profitability of the production company. In case of other parameters increase, the situation needs to be analysed further owing to the higher total cost. In order to keep similar total cost, management may consider suitable response in the form of any other appropriate parameter variation if it is possible. Such response is summarised in Table 1.3.

Table 1.3: Response with an objective of similar total cost.

Variation triggered by	*Management response*
Increase in setup cost	Holding cost reduction/ Production rate decrease
Increase in holding cost	Setup cost reduction/ Production rate decrease
Increase in production rate	Setup/ holding cost reduction

1.3.2 Case with shortages

When shortages are included in the production system, then the optimal production lot size and related annual total cost are obtained before. In case of the parameter variation, lot size and total cost will no longer remain similar. Depending on the need, the objective may be as follows:

(a) Similar lot size

(b) Similar total annual cost

Table 1.4 shows the potential management response concerning similar lot size.

Table 1.4: Response in case of shortages for similar lot size.

Variation triggered by	*Response*
Demand increase	Setup cost reduction/ Production rate increase
Demand decrease	Inventory holding cost reduction/ Shortage cost reduction/ Production rate decrease
Setup cost increase	Production rate increase
Setup cost decrease	Holding cost reduction/ Shortage cost reduction/ Production rate decrease/ Demand increase
Holding cost increase	Demand increase/ Shortage cost reduction/ Production rate decrease
Holding cost decrease	Setup cost reduction/ Production rate increase
Shortage cost increase	Demand increase/ Holding cost reduction/ Production rate decrease
Shortage cost reduction	Setup cost reduction/ Production rate increase
Production rate increase	Demand increase/ Holding cost reduction/ Shortage cost reduction
Production rate decrease	Setup cost reduction

With the inclusion of shortages, suitable response is summarised in Table 1.5 for an objective of similar total annual cost. Suitable response aims at reducing the total cost to the previous level if it increases by a parameter increase. Total cost decrease might not be a matter of concern corresponding to certain parameter variation particularly on downward side. However, in case of demand, an upward variation is always preferred in general because it contributes to the profitability by direct selling.

Table 1.5: Response in case of shortages for similar total cost.

Variation triggered by	*Response*
Increase in setup cost	Holding cost reduction/ Shortage cost reduction/ Production rate decrease
Increase in holding cost	Setup cost reduction/ Production rate decrease/ Shortage cost reduction
Increase in shortage cost	Setup cost reduction/ Production rate decrease/ Holding cost reduction
Increase in production rate	Setup/Holding/Shortage cost reduction

2
Customer Demand

Abstract : Prediction of customer demand depends on several aspects. After a brief discussion, a variation in demand has been narrated. In addition to a variety of reasons for an increased demand, the situation is further analysed. Influence of this demand increase on the output parameters such as batch quantity and total cost has been observed. Similarly after knowing the reasons for a decreased demand, the influence on the output parameters has been evaluated. Generalised results have also been tabulated after relevant derivations. Stock out cases are additionally described for both a higher as well as a lower customer demand.

Keyword : Demand variation, output parameters, higher demand, lower demand, generalised results

Customer demand is a significant parameter in an overall production system environment. Demand can be predicted and depends on several aspects, some of which are as follows:

(a) Country/region
(b) Population
(c) Age
(d) Habit
(e) Type and nature of product
(f) Season

Additionally, product demand estimation may depend on the industrial/ business sector such as FMCG, agri-products, and engineering products.

(i) FMCG:
In case of the fast-moving consumer goods, past consumption data from different regions can be consolidated in order to arrive at a total demand and a forecast can be created for the next planning period.

(ii) Agri-products:
In case of the agricultural products and its variants, demand may be seasonal also in certain cases. This factor also needs to be considered for the demand estimation, and after arriving at a suitable demand, batch size may be determined for furthermore processing of the agricultural/food products.

(iii) Engineering products:
In case of the engineering companies, most of the products are manufactured in batches at one or other stages of a series of production processes. Demands

of such products are also forecasted or calculated. Demand might be dependent on customer groups, consumers' purchasing power, and the need of the customer company.

Generally speaking also, demand is one of the significant parameters for industrial production. Depending on its estimation, optimal batch size is determined for the production planning. In a practical economic environment, demand can increase or decrease and its effects need to be understood properly.

2.1 Demand increase

Product demand can increase because of variety of reasons as shown in Fig. 2.1. Because of the comparative lower price, the customers may like to purchase more quantities resulting in a demand increase. Occasionally, price discount may be offered by the producer/supplier company resulting in a lower price for certain duration. By way of advertising at an appropriate time, demands might be increased. Sometimes because of higher income level of the target consumers, product demand may increase. In a particular season such as summer/winter, demand of certain type of products shows an upward trend. In case where an improved feature has been added in a product such as mobile telephone, demand may increase. Because of the changed business requirements, the demand of certain equipment/facility can be higher. On account of a focus on cashless transactions, more number of retailers may tend to buy the related facility. Demand of such facility producing company may become higher.

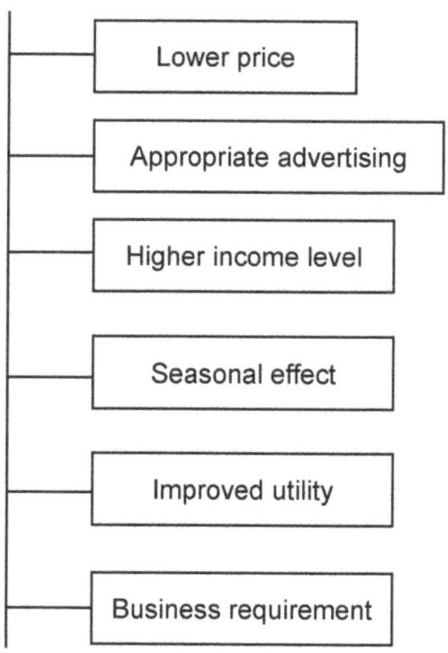

Figure 2.1: Variety of reasons for increased demand

With the increased demand (say $D1$), the production batch size increases, and facility is to be run for a longer period. This is shown in Fig. 2.2.

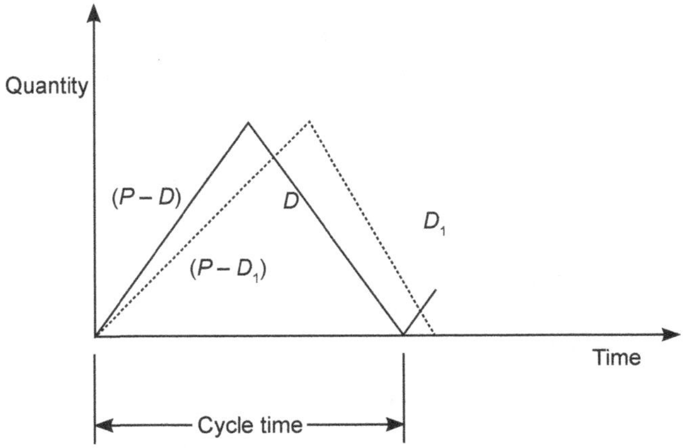

Figure 2.2: Demand increase

Example 2.1

Consider the following parameters:

Annual demand, $D = 600$ units

Setup cost, $C =$ ₹ 45

Annual production rate, $P = 960$ units

Annual inventory carrying cost per unit, $I =$ ₹ 40

Now:

From Eq.(1.6), production batch size:

$$Q^* = \sqrt{\frac{2DC}{(1 - D/P)I}}$$

$$= \sqrt{\frac{2 \times 600 \times 45}{(1 - 600/960) \times 40}}$$

$$= 60 \text{ units}$$

From Eq. (1.7), total annual cost:

$$E^* = \sqrt{2DCI(1 - D/P)}$$

$$= \sqrt{2 \times 600 \times 45 \times 40(1 - 600/960)}$$

$$= ₹\ 900$$

Now, if the demand is increased by 20% (say), then the revised annual demand is 720 units. The calculation for a new production lot size and total cost is as follows:

$$Q^* = \sqrt{\frac{2 \times 720 \times 45}{(1 - 720/960) \times 40}}$$

$$= 80.50 \text{ units}$$

And:

$$E^* = \sqrt{2 \times 720 \times 45 \times 40(1 - 720/960)}$$

$$= ₹\ 804.98$$

On comparison with the base parameter for demand:

Relative change in $Q = \dfrac{80.50 - 60}{60} = \dfrac{20.50}{60} = 0.3416$

% change in $Q = 3.416 \times 100 = 34.16\%$

Similarly % change in $E = \dfrac{900 - 804.98}{900} = \dfrac{95.02}{900} = 0.1056 = 10.56\%$

The production batch size has increased while the total cost is decreased. It is of interest to know the effects of increase in demand on the output parameters.

Example 2.2

Consider the input parameters of Example 2.1. Analyse if an increase in demand is as follows:

% Increase in D	5%	10%	15%	20%	25%	30%
D	630	660	690	720	750	780

The procedure has been detailed in Example 2.1. Following that, the variations in parameters with reference to a demand increase are shown in Table 2.1.

Table 2.1: Influence of demand increase on parameters.

% Increase in D	5%	10%	15%	20%	25%	30%
D	630	660	690	720	750	780
Q	64.22	68.93	74.30	80.50	87.83	96.75
% Increase in Q	7.03%	14.89%	23.83%	34.16%	46.39%	61.25%
E	882.96	861.68	835.84	804.98	768.52	725.60
% Decrease in E	1.89%	4.26%	7.13%	10.56%	14.61%	19.38%

Such results help in advance planning that pertains to production/manufacturing, transportation/material handling and purchase of input items contributing towards the related finished product. Refer Table 2.1. % variation in production batch size is much higher in comparison with that in the total cost.

In order to generalise, let:

M = % variation in parameter

In the present context, M relates to the % increase in demand, therefore the increased demand:

$$D_1 = \left(1 + \frac{M}{100}\right) D$$

Increase in the production batch size $= \sqrt{\dfrac{2D_1C}{(1 - D_1/P)I}} - \sqrt{\dfrac{2DC}{(1 - D/P)I}}$

$$= \sqrt{\frac{2DC}{(1 - D/P)I}}\left[\sqrt{\frac{D_1(1 - D/P)}{D(1 - D_1/P)}} - 1\right]$$

$$= \sqrt{\frac{2DC}{(1 - D/P)I}}\left[\sqrt{\frac{(1 + M/100)(1 - D/P)}{1 - (D/P)(1 + M/100)}} - 1\right]$$

And:

$$\% \text{ increase in } Q = \sqrt{\frac{(1 + M/100)(1 - D/P)}{1 - (D/P)(1 + M/100)}} - 1$$

Now:

Reduction in the total cost $= \sqrt{2DCI(1 - D/P)} - \sqrt{2D_1CI(1 - D_1/P)}$

$$= \sqrt{2DCI(1 - D/P)}\left[1 - \sqrt{\frac{D_1(1 - D_1/P)}{D(1 - D/P)}}\right]$$

$$= \sqrt{2DCI(1 - D/P)}\left[1 - \sqrt{\frac{(1 + M/100)\{1 - (D/P)(1 + M/100)\}}{(1 - D/P)}}\right]$$

And:

$$\% \text{ reduction in } E = 1 - \sqrt{\frac{(1 + M/100)\{1 - (D/P)(1 + M/100)\}}{(1 - D/P)}}$$

Table 2.2 summarises the derived results.

Table 2.2: Results with respect to % demand increase.

Increase in the production batch size	$\sqrt{\frac{2DC}{(1-D/P)I}}\left[\sqrt{\frac{(1+M/100)(1-D/P)}{1-(D/P)(1+M/100)}}-1\right]$
% increase in the production batch size	$\sqrt{\frac{(1+M/100)(1-D/P)}{1-(D/P)(1+M/100)}}-1$
Reduction in the total cost	$\sqrt{2DCI(1-D/P)}\left[1-\sqrt{\frac{(1+M/100)\{1-(D/P)(1+M/100)\}}{(1-D/P)}}\right]$
% reduction in the total cost	$1-\sqrt{\frac{(1+M/100)\{1-(D/P)(1+M/100)\}}{(1-D/P)}}$

2.2 Demand decrease

Demand might decrease because of the following reasons (Fig. 2.3):

(i) Product has comparatively higher price and therefore the customers are not coming forward.

(ii) There is a competitor product available in the market and therefore demand is at a decreased level probably because of superiority of competitor product.

(iii) The target consumer group has certain financial level which is not corresponding to the desired purchasing power for a product.

(iv) Demand might go down since there is no season.

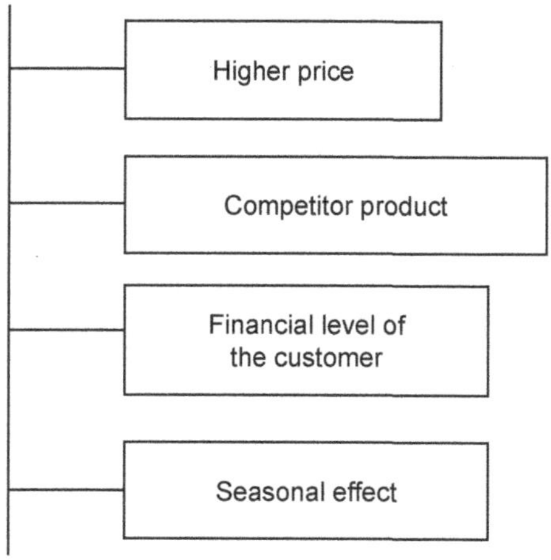

Figure 2.3: Variety of reasons for decreased demand

With the decreased demand (say *D*1), the production batch size reduces and facility is to be run for shorter period. Fig. 2.4 represents this case.

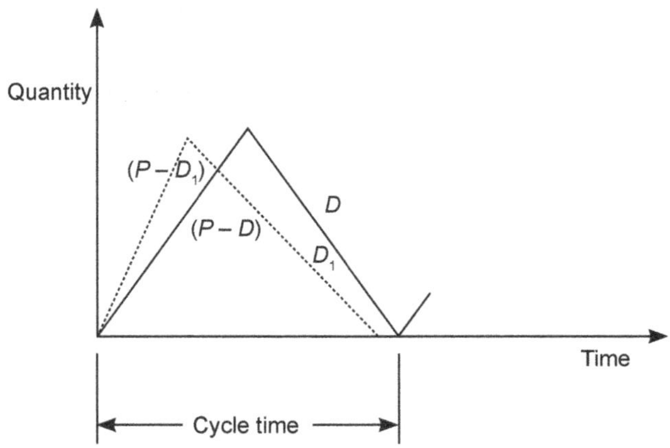

Figure 2.4: Demand decrease

Example 2.3

Consider the input parameters of Example 2.1. Analyse if a reduction in *D* is as follows:

% Decrease in D	5%	10%	15%	20%	25%	30%
D	570	540	510	480	450	420

The variations in parameters with reference to a demand decrease are represented by Table 2.3.

Table 2.3: Influence of demand decrease on parameters.

% Decrease in D	5%	10%	15%	20%	25%	30%
D	570	540	510	480	450	420
Q	56.187	52.699	49.477	46.476	43.656	40.988
% Decrease in Q	6.36%	12.17%	17.54%	22.54%	27.24%	31.69%
E	913.03	922.23	927.70	929.52	927.70	922.23
% Increase in E	1.45%	2.47%	3.08%	3.28%	3.08%	2.47%

Refer Table 2.3. % variation in production batch size is much higher in comparison with that in the total cost. With a reduction in demand, production batch size reduces and total cost increases.

In order to generalise, let:

M = % variation in parameter

In the present context, M relates to the % decrease in demand, therefore the decreased demand:

$$D_1 = \left(1 - \frac{M}{100}\right) D$$

Decrease in the production batch size $= \sqrt{\frac{2DC}{(1-D/P)I}} - \sqrt{\frac{2D_1C}{(1-D_1/P)I}}$

$$= \sqrt{\frac{2DC}{(1-D/P)I}}\left[1 - \sqrt{\frac{D_1(1-D/P)}{D(1-D_1/P)}}\right]$$

$$= \sqrt{\frac{2DC}{(1-D/P)I}}\left[1 - \sqrt{\frac{(1-M/100)(1-D/P)}{1-(D/P)(1-M/100)}}\right]$$

And:

% decrease in $Q = 1 - \sqrt{\frac{(1-M/100)(1-D/P)}{1-(D/P)(1-M/100)}}$

Now:

Increase in the total cost $= \sqrt{2D_1CI(1-D_1/P)} - \sqrt{2DCI(1-D/P)}$

$$= \sqrt{2DCI(1-D/P)}\left[\sqrt{\frac{D_1(1-D_1/P)}{D(1-D/P)}} - 1\right]$$

$$= \sqrt{2DCI(1-D/P)}\left[\sqrt{\frac{(1-M/100)\{1-(D/P)(1-M/100)\}}{(1-D/P)}} - 1\right]$$

And:

% increase in $E = \sqrt{\frac{(1-M/100)\{1-(D/P)(1-M/100)\}}{(1-D/P)}} - 1$

Table 2.4 summarises the derived results.

Table 2.4: Results with respect to % demand decrease.

Decrease in the production batch size	$\sqrt{\frac{2DC}{(1-D/P)I}}\left[1-\sqrt{\frac{(1-M/100)(1-D/P)}{1-(D/P)(1-M/100)}}\right]$
% decrease in the production batch size	$1-\sqrt{\frac{(1-M/100)(1-D/P)}{1-(D/P)(1-M/100)}}$

Contd...

Contd...

Increase in the total cost	$\sqrt{2DCI(1-D/P)}\left[\sqrt{\dfrac{(1-M/100)\{1-(D/P)(1-M/100)\}}{(1-D/P)}}-1\right]$
% increase in the total cost	$\sqrt{\dfrac{(1-M/100)\{1-(D/P)(1-M/100)\}}{(1-D/P)}}-1$

Observe Tables 2.1 and 2.3. % variation in manufacturing batch size is more in case of the demand increase case. Variation in total cost is also higher in case of the demand increase.

The present analysis will be helpful in determining the production or manufacturing batch size and a related total cost for the following:

(a) Existence of certain level of demand for any product

(b) Increase in demand owing to few factors

(c) Demand reduction because of some reasons

A trend is made available with reference to the fluctuation in demand. Storage and manufacturing space may be better utilised/arranged depending on the information such as additional space requirement or extra space availability. In case where manufacturing batch size increases, the need for additional space arises and it requires to be planned in the context of:

(i) Area available for storage of the final product

(ii) Intermediate area available between facilities

In case where production lot size reduces, the extra space is available to the management. Utilisation of extra space can be better planned if the quantitative information is available beforehand.

In a situation when the related total cost increases, it needs to be considered additionally well in advance in the context of:

(i) Available money

(ii) Workforce/facility utilisation

(iii) The effect on profitability

When the related total cost decreases, the current level of money availability should be observed. Resources need to be balanced in terms of workforce or facility. Such aspects may be planned well in advance and the effect on profit level can be analysed better.

However, the discussed analysis does not include stock out cases. When shortages or stock outs are allowed in the production system, computational effects must be determined and also generalised results should be derived.

2.3 Stock out case

In case of the stock out or shortages, the optimum values for output parameters have been obtained by the Eqs. (1.16), (1.17), and (1.18).

Example 2.4

Consider the following parameters:

Annual demand, $D = 600$ units

Setup cost, $C =$ ₹ 45

Annual production rate, $P = 960$ units

Annual inventory carrying cost per unit, $I =$ ₹ 40

Annual shortage cost per unit, $K =$ ₹ 100

Using the Eq. (1.16),

$$Q^* = \sqrt{\frac{2DC(K+I)}{KI(1-D/P)}}$$

$$= 71 \text{ units}$$

From the Eq. (1.17), an optimum shortage quantity:

$$J^* = \sqrt{\frac{2DCI(1-D/P)}{K(K+I)}}$$

$$= 7.61 \text{ units}$$

And the total annual cost from Eq. (1.18),

$$E^* = \sqrt{\frac{2DCKI(1-D/P)}{(K+I)}}$$

$$= ₹\ 760.64$$

These results along with the input parameters are also shown below:

P	D	C	I	K	Q	E	J
960	600	45	40	100	71.0	760.64	7.606388

On comparison with Example 2.1, production batch size has increased along with an overall reduction in the total cost. This is because:

$$\sqrt{\frac{2DC(K+I)}{KI(1-D/P)}} > \sqrt{\frac{2DC}{(1-D/P)I}}$$

Or $$\frac{(K+I)}{K} > 1$$

That is true.

And:

$$\sqrt{\frac{2DCKI(1-D/P)}{(K+I)}} < \sqrt{2DCI(1-D/P)}$$

Or

$$\frac{K}{(K+I)} < 1$$

That is also true.

With reference to the inclusion of stock out, the following output parameters have been obtained:

(i) Batch size
(ii) Shortage quantity
(iii) Related total cost

While comparing with the case without stock outs, the effects are as follows:

(a) Increased production batch size is observed to suitably incorporate the shortages or stock outs.
(b) An overall reduced total cost is observed while incorporating the shortage quantities.

Along with the precise estimation of shortage cost, this benefit of reduced total cost can be achieved in appropriate situations.

2.3.1 Higher demand

When demand is higher, i.e., increased from D to D_1, Fig. 2.5 illustrates this situation (keeping the production rate similar).

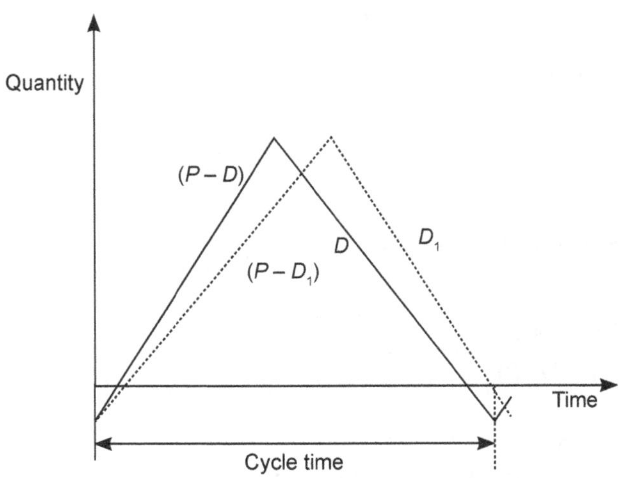

Figure 2.5: Higher demand with shortages

Example 2.5

Consider the information from previous Example:

P	*D*	*C*	*I*	*K*	*Q*	*E*	*J*
960	600	45	40	100	71.0	760.64	7.606388

Implement the higher demand as follows:

% Increase in D	5%	10%	15%	20%	25%	30%
D	630	660	690	720	750	780

Following the procedure explained in the previous Example, the results are shown in Table 2.5. The production batch size increases along with reduction in the total cost. Maximum shortage quantities decrease. However, % reduction in J is similar to that in E.

Table 2.5: Influence of demand increase on parameters along with shortages.

% Increase in D	5%	10%	15%	20%	25%	30%
D	630	660	690	720	750	780
Q	75.981	81.565	87.909	95.247	103.923	114.473
% Increase in Q	7.03%	14.89%	23.83%	34.16%	46.39%	61.25%
E	746.24	728.26	706.41	680.34	649.52	613.25
% Decrease in E	1.89%	4.26%	7.13%	10.56%	14.61%	19.38%
J	7.46	7.28	7.06	6.80	6.50	6.13
% Decrease in J	1.89%	4.26%	7.13%	10.56%	14.61%	19.38%

For a general approach:

$$D_1 = \left(1+\frac{M}{100}\right)D$$

Increase in the production batch size $= \sqrt{\frac{2D_1C(K+I)}{KI(1-D_1/P)}} - \sqrt{\frac{2DC(K+I)}{KI(1-D/P)}}$

$$= \sqrt{\frac{2DC(K+I)}{KI(1-D/P)}}\left[\sqrt{\frac{D_1(1-D/P)}{D(1-D_1/P)}} - 1\right]$$

$$= \sqrt{\frac{2DC(K+I)}{KI(1-D/P)}}\left[\sqrt{\frac{(1+M/100)(1-D/P)}{1-(D/P)(1+M/100)}} - 1\right]$$

And:

$$\% \text{ increase in } Q = \sqrt{\frac{(1+M/100)(1-D/P)}{1-(D/P)(1+M/100)}} - 1$$

Now:

$$\text{Reduction in the total cost} = \sqrt{\frac{2DCKI(1-D/P)}{(K+I)}} - \sqrt{\frac{2D_1CKI(1-D_1/P)}{(K+I)}}$$

$$= \sqrt{\frac{2DCKI(1-D/P)}{(K+I)}}\left[1-\sqrt{\frac{D_1(1-D_1/P)}{D(1-D/P)}}\right]$$

$$= \sqrt{\frac{2DCKI(1-D/P)}{(K+I)}}\left[1-\sqrt{\frac{(1+M/100)\{1-(D/P)(1+M/100)\}}{(1-D/P)}}\right]$$

And:

$$\% \text{ reduction in } E = 1-\sqrt{\frac{(1+M/100)\{1-(D/P)(1+M/100)\}}{(1-D/P)}}$$

Similarly:
Reduction in the maximum shortage quantity

$$= \sqrt{\frac{2DCI(1-D/P)}{K(K+I)}} - \sqrt{\frac{2D_1CI(1-D_1/P)}{K(K+I)}}$$

$$= \sqrt{\frac{2DCI(1-D/P)}{K(K+I)}}\left[1-\sqrt{\frac{D_1(1-D_1/P)}{D(1-D/P)}}\right]$$

$$= \sqrt{\frac{2DCI(1-D/P)}{K(K+I)}}\left[1-\sqrt{\frac{(1+M/100)\{1-(D/P)(1+M/100)\}}{(1-D/P)}}\right]$$

And:

$$\% \text{ reduction in } J = 1-\sqrt{\frac{(1+M/100)\{1-(D/P)(1+M/100)\}}{(1-D/P)}}$$

Table 2.6 summarises the derived results.

Table 2.6: Results with respect to % demand increase including shortages.

Increase in the production batch size	$\sqrt{\frac{2DC(K+I)}{KI(1-D/P)}}\left[\sqrt{\frac{(1+M/100)(1-D/P)}{1-(D/P)(1+M/100)}}-1\right]$
% increase in the production batch size	$\sqrt{\frac{(1+M/100)(1-D/P)}{1-(D/P)(1+M/100)}}-1$
Reduction in the total cost	$\sqrt{\frac{2DCKI(1-D/P)}{(K+I)}}\left[1-\sqrt{\frac{(1+M/100)\{1-(D/P)(1+M/100)\}}{(1-D/P)}}\right]$
% reduction in the total cost	$1-\sqrt{\frac{(1+M/100)\{1-(D/P)(1+M/100)\}}{(1-D/P)}}$
Reduction in the maximum shortage quantity	$\sqrt{\frac{2DCI(1-D/P)}{K(K+I)}}\left[1-\sqrt{\frac{(1+M/100)\{1-(D/P)(1+M/100)\}}{(1-D/P)}}\right]$
% reduction in J	$1-\sqrt{\frac{(1+M/100)\{1-(D/P)(1+M/100)\}}{(1-D/P)}}$

After observing the effects of higher demand on output parameters, the generalised results are presented. With higher demand, the behavior of output parameters is given as follows:

(i) Higher production batch size

(ii) Lower total related cost

(iii) Lower optimal stock out quantity

Managerial implications of higher production batch size and lower total cost have been discussed in previous scenario, i.e., without stock outs. Implementation of lower optimal stock out quantity results into relatively less potential customer dissatisfaction.

2.3.2 Lower demand:

When demand is lowered from D to $D1$, it is depicted in Fig. 2.6 (keeping the production rate similar).

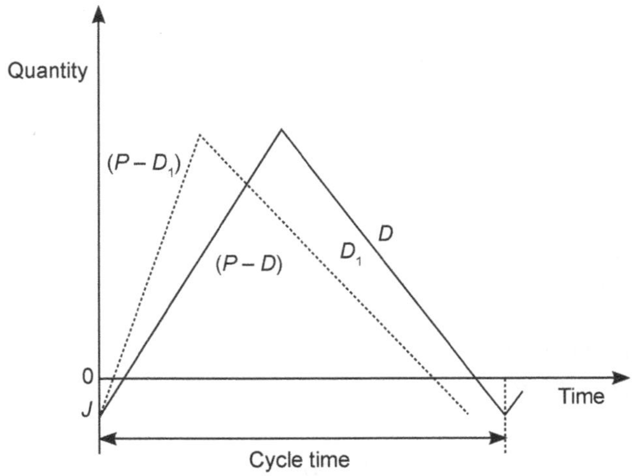

Figure 2.6: Lower demand with shortages

Example 2.6

Consider the base data with an inclusion of shortages as follows:

P	*D*	*C*	*I*	*K*	*Q*	*E*	*J*
960	600	45	40	100	71.0	760.64	7.606388

Implement the lower demand as follows:

% Decrease in D	5%	10%	15%	20%	25%	30%
D	570	540	510	480	450	420

The computational results are shown in Table 2.7. The production batch size reduces along with an increase in total cost. Maximum shortage quantity also increases. However, % increase in it is similar to % increase in total cost.

Table 2.7: Influence of demand decrease on parameters along with shortages.

% Decrease in D	5%	10%	15%	20%	25%	30%
D	570	540	510	480	450	420
Q	66.481	62.354	58.542	54.991	51.655	48.497
% decrease in Q	6.36%	12.17%	17.54%	22.54%	27.24%	31.69%
E	771.65	779.42	784.05	785.58	784.05	779.42
% Increase in E	1.45%	2.47%	3.08%	3.28%	3.08%	2.47%
J	7.72	7.79	7.84	7.86	7.84	7.79
% Increase in J	1.45%	2.47%	3.08%	3.28%	3.08%	2.47%

For a general approach:

$$D_1 = \left(1 - \frac{M}{100}\right) D$$

Decrease in the production batch size $= \sqrt{\frac{2DC(K+I)}{KI(1-D/P)}} - \sqrt{\frac{2D_1C(K+I)}{KI(1-D_1/P)}}$

$$= \sqrt{\frac{2DC(K+I)}{KI(1-D/P)}}\left[1 - \sqrt{\frac{D_1(1-D/P)}{D(1-D_1/P)}}\right]$$

$$= \sqrt{\frac{2DC(K+I)}{KI(1-D/P)}}\left[1 - \sqrt{\frac{(1-M/100)(1-D/P)}{1-(D/P)(1-M/100)}}\right]$$

And:

% decrease in $Q = 1 - \sqrt{\frac{(1-M/100)(1-D/P)}{1-(D/P)(1-M/100)}}$

Now:

Increase in the total cost $= \sqrt{\frac{2D_1CKI(1-D_1/P)}{(K+I)}} - \sqrt{\frac{2DCKI(1-D/P)}{(K+I)}}$

$$= \sqrt{\frac{2DCKI(1-D/P)}{(K+I)}}\left[\sqrt{\frac{D_1(1-D_1/P)}{D(1-D/P)}} - 1\right]$$

$$= \sqrt{\frac{2DCKI(1-D/P)}{(K+I)}}\left[\sqrt{\frac{(1-M/100)\{1-(D/P)(1-M/100)\}}{(1-D/P)}} - 1\right]$$

And:

% increase in $E = \sqrt{\frac{(1-M/100)\{1-(D/P)(1-M/100)\}}{(1-D/P)}} - 1$

Similarly:

Increase in the maximum shortage quantity

$$= \sqrt{\frac{2D_1CI(1-D_1/P)}{K(K+I)}} - \sqrt{\frac{2DCI(1-D/P)}{K(K+I)}}$$

$$= \sqrt{\frac{2DCI(1-D/P)}{K(K+I)}}\left[\sqrt{\frac{D_1(1-D_1/P)}{D(1-D/P)}} - 1\right]$$

$$= \sqrt{\frac{2DCI(1-D/P)}{K(K+I)}}\left[\sqrt{\frac{(1-M/100)\{1-(D/P)(1-M/100)\}}{(1-D/P)}} - 1\right]$$

And:

$$\% \text{ increase in } J = \sqrt{\frac{(1-M/100)\{1-(D/P)(1-M/100)\}}{(1-D/P)}} - 1$$

Table 2.8 summarises the derived results.

Table 2.8: Results with respect to % demand decrease including shortages.

Decrease in the production batch size	$\sqrt{\frac{2DC(K+I)}{KI(1-D/P)}}\left[1-\sqrt{\frac{(1-M/100)(1-D/P)}{1-(D/P)(1-M/100)}}\right]$
% decrease in the production batch size	$1-\sqrt{\frac{(1-M/100)(1-D/P)}{1-(D/P)(1-M/100)}}$
Increase in the total cost	$\sqrt{\frac{2DCKI(1-D/P)}{(K+I)}}\left[\sqrt{\frac{(1-M/100)\{1-(D/P)(1-M/100)\}}{(1-D/P)}}-1\right]$
% increase in the total cost	$\sqrt{\frac{(1-M/100)\{1-(D/P)(1-M/100)\}}{(1-D/P)}}-1$
Increase in the maximum shortage quantity	$\sqrt{\frac{2DCI(1-D/P)}{K(K+I)}}\left[\sqrt{\frac{(1-M/100)\{1-(D/P)(1-M/100)\}}{(1-D/P)}}-1\right]$
% increase in J	$\sqrt{\frac{(1-M/100)\{1-(D/P)(1-M/100)\}}{(1-D/P)}}-1$

With lower demand, an implementation of derived output parameters results in:

(a) Lower production batch size

(b) Higher optimal stock out quantity

Additionally, the total related cost might be higher. However, it might be analysed in practice along with a change in the procurement requirement in addition to the present production environment alone. Lower manufacturing batch size may result in availability of extra space at the finished product level as well as at the work-in-process inventory level, i.e., in the context of intermediate space between facilities. Higher level of stock out might lead towards a suitable assessment of its effects in the business.

3

Setup Cost

Abstract : Factors affecting setup and the setup cost estimation have been mentioned in this chapter. The reasons for a lower and a higher setup cost are discussed. Examples are provided for a study of a downward and an upward variation in the facility setup cost. Effects of such variations in setup cost have been studied in the context of related parameters. With an inclusion of stock outs, a maximum shortage quantity is also analysed along with the production batch size and the total annual cost. An interaction of demand and setup cost is especially investigated with an objective of similar production batch size.

Keyword : Setup cost estimation, lower setup cost, higher setup cost, interaction of demand and setup cost, similar production batch size

A production facility should be setup properly before starting any actual manufacturing activity.

3.1 Setup cost estimation

In order to setup a manufacturing facility, some of the activities are as follows:

(i) The machine/facility needs to be prepared in order to undertake any desired process including a general cleaning.

(ii) Job holders including fixtures need to be arranged and aligned.

(iii) Tools and tool holders need to be arranged and aligned.

(iv) Significant production process parameters should be set on the facility/machine. For example, speed, feed, and depth of cut should be ascertained and finally implemented on the machine in case of turning operation. In case of welding process, voltage and current need to be set.

(v) Before actual start of the production, trial run may be necessary in some cases. This might include wastage of few pieces of component.

These activities involve time and efforts. Costs including salary/wages of the engineer/worker/employee need to be estimated among other aspects in order to finally arrive at suitable setup cost for one setup concerning a facility. This may be treated as fixed setup cost per setup.

Additionally setup and associated cost depend on few factors as shown in Fig. 3.1.

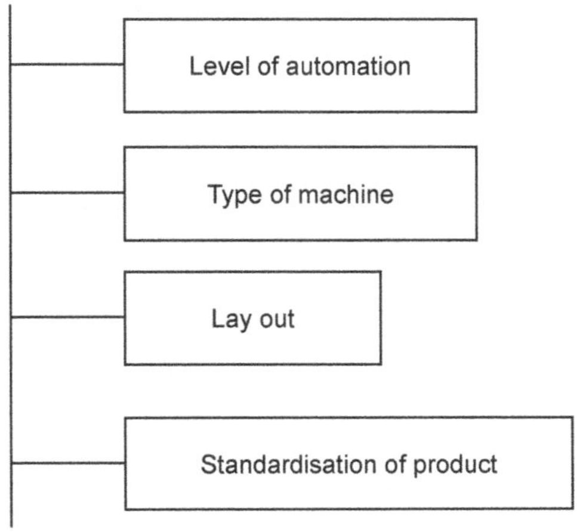

Figure 3.1: Factors affecting setup

If level of automation is higher, the time and efforts concerning setup are less making setup cost lower. Type of the machine such as general purpose or special purpose also has certain influence on setup issues. General purpose machine can be suitable for a wide variety of products with major setup efforts and time. With minor effort, special purpose machine is ready for manufacture of specific product and associated variants in similar category. Layout of the factory also affects the time and activities related to setup. For example, if the tool room is far from the location of machine, then there is a possibility of more handling of the related accessories/tools. This may result in more setup cost. In case where product is standard, the requirement for calculation of various parameters is of lower degree and time consumed is less. Furthermore, tools/dies or jigs/fixtures might also be standard and preparation might take less time thus saving in resources.

3.2 Lower setup cost

Setup cost may be lower because of the following reasons:

(a) The concerned employees have got certain experience on the machine and therefore time taken in setup is now reduced.

(b) A senior worker had a major role in setup, who is now retired. Now another young employee does that job whose salary level is comparatively less. Therefore, the setup cost is now lower.

Effects of lower setup cost on output parameters are important and useful for furthermore planning activity accordingly.

Example 3.1

With the following parameters:

Annual demand, $D = 600$ units

Setup cost, $C = ₹\ 45$

Annual production rate, $P = 960$ units

Annual inventory carrying cost per unit, $I = ₹\ 40$;

Production batch size:

$$Q^* = \sqrt{\frac{2DC}{(1 - D/P)I}} = 60 \text{ units}$$

And total annual cost:

$$E^* = \sqrt{2DCI(1 - D/P)} = ₹\ 900$$

Parameters for this basic case are as follows:

D	*C*	*I*	*P*	*Q*	*E*
600	45	40	960	60.00	900.00

Analyse the reduction in setup cost as follows:

% Decrease in C	5%	10%	15%	20%	25%	30%
C	42.75	40.5	38.25	36	33.75	31.5

The effects of lower setup cost are provided in Table 3.1. It has resulted into:

(i) Lower production batch size

(ii) Lower overall total annual cost

Percentage variations in both these parameters are similar.

Table 3.1: Influence of lower setup cost on parameters.

% Decrease in C	5%	10%	15%	20%	25%	30%
C	42.75	40.5	38.25	36	33.75	31.5
Q	58.481	56.921	55.317	53.666	51.962	50.200
% Decrease in Q	2.53%	5.13%	7.80%	10.56%	13.40%	16.33%
E	877.21	853.81	829.76	804.98	779.42	752.99
% Decrease in E	2.53%	5.13%	7.80%	10.56%	13.40%	16.33%

For a generalisation, let:

M = % reduction in setup cost

Now:

$$C_1 = \left(1-\frac{M}{100}\right)C$$

Reduction in the production batch size = $\sqrt{\frac{2DC}{(1-D/P)I}} - \sqrt{\frac{2DC_1}{(1-D/P)I}}$

$$= \sqrt{\frac{2DC}{(1-D/P)I}}\left[1-\sqrt{\frac{C_1}{C}}\right]$$

$$= \sqrt{\frac{2DC}{(1-D/P)I}}\left[1-\sqrt{1-\frac{M}{100}}\right]$$

And:

% reduction in $Q = 1-\sqrt{1-\frac{M}{100}}$

Now:

Reduction in the total cost = $\sqrt{2DCI(1-D/P)} - \sqrt{2DC_1I(1-D/P)}$

$$= \sqrt{2DCI(1-D/P)}\left[1-\sqrt{\frac{C_1}{C}}\right]$$

$$= \sqrt{2DCI(1-D/P)}\left[1-\sqrt{1-\frac{M}{100}}\right]$$

And:

% reduction in $E = 1-\sqrt{1-\frac{M}{100}}$

Table 3.2 shows the derived results.

Table 3.2: Results with respect to % setup cost reduction.

Reduction in the production batch size	$\sqrt{\frac{2DC}{(1-D/P)I}}\left[1-\sqrt{1-\frac{M}{100}}\right]$
% reduction in the production batch size	$1-\sqrt{1-\frac{M}{100}}$
Reduction in the total cost	$\sqrt{2DCI(1-D/P)}\left[1-\sqrt{1-\frac{M}{100}}\right]$
% reduction in the total cost	$1-\sqrt{1-\frac{M}{100}}$

3.3 Higher setup cost

Setup cost may be higher because of the following reasons:

(a) Electricity charges have gone up and therefore expenses involved in setting up of the machine may also go up. For the trial run before actual production, machine consumes power and therefore power charges might have certain influence on setup cost estimation.

(b) After certain experience, employee salary goes up affecting setup cost upward. Effects of higher setup cost on output parameters are important and useful for furthermore planning activity accordingly.

Example 3.2

With the data from previous Example, as follows:

D	*C*	*I*	*P*	*Q*	*E*
600	45	40	960	60.00	900.00

Analysis is conducted with respect to higher setup cost as follows:

% Increase in C	5%	10%	15%	20%	25%	30%
C	47.25	49.5	51.75	54	56.25	58.5

Computational results are provided in Table 3.3. Both the production batch size and total cost have increased with similar % variation. Because of an increased batch size, storage space needs to be arranged including space between facilities as the work-in-process inventory related to manufacturing activity.

Table 3.3: Influence of higher setup cost on parameters.

% Increase in C	5%	10%	15%	20%	25%	30%
C	47.25	49.5	51.75	54	56.25	58.5
Q	61.482	62.929	64.343	65.727	67.082	68.411
% Increase in Q	2.47%	4.88%	7.24%	9.54%	11.80%	14.02%
E	922.23	943.93	965.14	985.90	1006.23	1026.16
% Increase in E	2.47%	4.88%	7.24%	9.54%	11.80%	14.02%

In order to generalise, let:

M = % variation in parameter

In the present context, M relates to the % increase in setup cost; therefore, the increased setup cost:

$$C_1 = \left(1+\frac{M}{100}\right)C$$

Increase in the production batch size = $\sqrt{\frac{2DC_1}{(1-D/P)I}} - \sqrt{\frac{2DC}{(1-D/P)I}}$

$$= \sqrt{\frac{2DC}{(1-D/P)I}}\left[\sqrt{\frac{C_1}{C}}-1\right]$$

$$= \sqrt{\frac{2DC}{(1-D/P)I}}\left[\sqrt{1+\frac{M}{100}}-1\right]$$

And:

% increase in $Q = \sqrt{1+\frac{M}{100}}-1$

Now:

Increase in the total cost = $\sqrt{2DC_1I(1-D/P)} - \sqrt{2DCI(1-D/P)}$

$$= \sqrt{2DCI(1-D/P)}\left[\sqrt{\frac{C_1}{C}}-1\right]$$

$$= \sqrt{2DCI(1-D/P)}\left[\sqrt{1+\frac{M}{100}}-1\right]$$

And:

% increase in $E = \sqrt{1+\frac{M}{100}}-1$

Table 3.4 shows the derived results.

Table 3.4: Results with respect to % setup cost increase.

Increase in the production batch size	$\sqrt{\frac{2DC}{(1-D/P)I}}\left[\sqrt{1+\frac{M}{100}}-1\right]$
% increase in the production batch size	$\sqrt{1+\frac{M}{100}}-1$
Increase in the total cost	$\sqrt{2DCI(1-D/P)}\left[\sqrt{1+\frac{M}{100}}-1\right]$
% increase in the total cost	$\sqrt{1+\frac{M}{100}}-1$

Compare Table 3.1 and 3.3% variations in output parameters are more in case of lower setup cost. This is because:

$$1-\sqrt{1-\frac{M}{100}} > \sqrt{1+\frac{M}{100}}-1$$

Or $$2 > \sqrt{1+\frac{M}{100}}+\sqrt{1-\frac{M}{100}}$$

Or $$2^2 > \left[\sqrt{1+\frac{M}{100}}+\sqrt{1-\frac{M}{100}}\right]^2$$

Or $$4 > 1+\frac{M}{100}+1-\frac{M}{100}+2\sqrt{1-\left(\frac{M}{100}\right)^2}$$

Or $$4 > 2+2\sqrt{1-\left(\frac{M}{100}\right)^2}$$

Or $$2 > 2\sqrt{1-\left(\frac{M}{100}\right)^2}$$

Or $$1 > \sqrt{1-\left(\frac{M}{100}\right)^2}$$

Or $$1 > 1-\left(\frac{M}{100}\right)^2$$

And this is true because the practical value of M is greater than zero.

3.4 Inclusion of stock outs

When shortages are incorporated in the production system, influence of setup cost variation is studied in this section.

Example 3.3

Consider the parameters of Example 3.1 with additional value of:

Annual shortage cost per unit, K = ₹ 100

Using the Eq. (1.16),

$$Q^* = \sqrt{\frac{2DC(K+I)}{KI(1-D/P)}}$$

=71 units

From the Eq. (1.17), an optimum shortage quantity:

$$J^* = \sqrt{\frac{2DCI(1-D/P)}{K(K+I)}}$$

$$= 7.61 \text{ units}$$

And the total annual cost from Eq. (1.18),

$$E^* = \sqrt{\frac{2DCKI(1-D/P)}{(K+I)}}$$

$$= ₹\ 760.64$$

3.4.1 Setup cost reduction

With setup cost reduction, the following parameters decrease:

(i) Production batch size

(ii) Maximum shortage quantity

(iii) Total annual cost

Example 3.4

Consider the information from previous Example:

P	*D*	*C*	*I*	*K*	*Q*	*E*	*J*
960	600	45	40	100	71.0	760.64	7.606388

Analyse the reduced setup cost as follows:

% Decrease in C	5%	10%	15%	20%	25%	30%
C	42.75	40.5	38.25	36	33.75	31.5

The computational results are shown in Table 3.5. As mentioned before, output parameters reduce. However, % variations in output parameters are similar.

Table 3.5: Influence of setup cost reduction on parameters along with shortages.

% Decrease in C	5%	10%	15%	20%	25%	30%
C	42.75	40.5	38.25	36	33.75	31.5
Q	69.195	67.350	65.452	63.498	61.482	59.397
% Decrease in Q	2.53%	5.13%	7.80%	10.56%	13.40%	16.33%
E	741.38	721.61	701.27	680.34	658.73	636.40
% Decrease in E	2.53%	5.13%	7.80%	10.56%	13.40%	16.33%
J	7.41	7.22	7.01	6.80	6.59	6.36
% Decrease in J	2.53%	5.13%	7.80%	10.56%	13.40%	16.33%

In order to generalise:

$$C_1 = \left(1 - \frac{M}{100}\right)C$$

Decrease in the production batch size $= \sqrt{\frac{2DC(K+I)}{KI(1-D/P)}} - \sqrt{\frac{2DC_1(K+I)}{KI(1-D/P)}}$

$$= \sqrt{\frac{2DC(K+I)}{KI(1-D/P)}}\left[1 - \sqrt{\frac{C_1}{C}}\right]$$

$$= \sqrt{\frac{2DC(K+I)}{KI(1-D/P)}}\left[1 - \sqrt{1 - \frac{M}{100}}\right]$$

And:

% decrease in $Q = 1 - \sqrt{1 - \frac{M}{100}}$

Now:

Decrease in the total cost $= \sqrt{\frac{2DCKI(1-D/P)}{(K+I)}} - \sqrt{\frac{2DC_1KI(1-D/P)}{(K+I)}}$

$$= \sqrt{\frac{2DCKI(1-D/P)}{(K+I)}}\left[1 - \sqrt{\frac{C_1}{C}}\right]$$

$$= \sqrt{\frac{2DCKI(1-D/P)}{(K+I)}}\left[1 - \sqrt{1 - \frac{M}{100}}\right]$$

And:

% decrease in $E = 1 - \sqrt{1 - \frac{M}{100}}$

Similarly, decrease in the maximum shortage quantity is

$$= \sqrt{\frac{2DCI(1-D/P)}{K(K+I)}} - \sqrt{\frac{2DC_1I(1-D/P)}{K(K+I)}}$$

$$= \sqrt{\frac{2DCI(1-D/P)}{K(K+I)}}\left[1 - \sqrt{\frac{C_1}{C}}\right]$$

$$= \sqrt{\frac{2DCI(1-D/P)}{K(K+I)}}\left[1 - \sqrt{1 - \frac{M}{100}}\right]$$

And:

$$\% \text{ decrease in } J = 1 - \sqrt{1 - \frac{M}{100}}$$

Table 3.6 shows the derived results. When shortages are incorporated in the system, these generalised results will help in precise calculation and analysis. These include the following:

(i) As production batch size reduces, more space is available relatively and it might be planned to utilise in a suitable way.

(ii) As the total expenditure is relatively less, the savings might be utilised in alternate ways if it is feasible.

(iii) As there is reduction in shortage quantity, the customers might be more satisfied in general in certain situations.

Table 3.6: Results with respect to % setup cost reduction including shortages.

Reduction in the production batch size	$\sqrt{\frac{2DC(K+I)}{KI(1-D/P)}}\left[1-\sqrt{1-\frac{M}{100}}\right]$
% reduction in the production batch size	$1-\sqrt{1-\frac{M}{100}}$
Reduction in the total cost	$\sqrt{\frac{2DCKI(1-D/P)}{(K+I)}}\left[1-\sqrt{1-\frac{M}{100}}\right]$
% reduction in the total cost	$1-\sqrt{1-\frac{M}{100}}$
Reduction in the maximum shortage quantity	$\sqrt{\frac{2DCI(1-D/P)}{K(K+I)}}\left[1-\sqrt{1-\frac{M}{100}}\right]$
% reduction in J	$1-\sqrt{1-\frac{M}{100}}$

3.4.2 Setup cost increase

With setup cost increase, the following parameters increase:

(i) Production batch size

(ii) Maximum shortage quantity

(iii) Total annual cost

Example 3.5

With the following information:

P	*D*	*C*	*I*	*K*	*Q*	*E*	*J*
960	600	45	40	100	71.0	760.64	7.606388

Implement the increased setup cost as follows:

% Increase in C	5%	10%	15%	20%	25%	30%
C	47.25	49.5	51.75	54	56.25	58.5

The results are shown in Table 3.7. As mentioned before, output parameters increase. However, % variations in output parameters are similar. Because of an increase in output parameters, the company should be able to plan for additional resources such as storage space and related expenditure if it is relevant. An increased total cost requirements should also be fulfilled by the manufacturing company.

Table 3.7: Influence of setup cost increase on parameters along with shortages.

% Increase in C	5%	10%	15%	20%	25%	30%
C	47.25	49.5	51.75	54	56.25	58.5
Q	72.746	74.458	76.131	77.769	79.373	80.944
% Increase in Q	2.47%	4.88%	7.24%	9.54%	11.80%	14.02%
E	779.42	797.76	815.69	833.24	850.42	867.26
% Increase in E	2.47%	4.88%	7.24%	9.54%	11.80%	14.02%
J	7.79	7.98	8.16	8.33	8.50	8.67
% Increase in J	2.47%	4.88%	7.24%	9.54%	11.80%	14.02%

In order to generalise:

$$C_1 = \left(1+\frac{M}{100}\right)C$$

$$\text{Increase in the production batch size} = \sqrt{\frac{2DC_1(K+I)}{KI(1-D/P)}} - \sqrt{\frac{2DC(K+I)}{KI(1-D/P)}}$$

$$= \sqrt{\frac{2DC(K+I)}{KI(1-D/P)}}\left[\sqrt{\frac{C_1}{C}}-1\right]$$

$$= \sqrt{\frac{2DC(K+I)}{KI(1-D/P)}}\left[\sqrt{1+\frac{M}{100}}-1\right]$$

And: % increase in $Q = \sqrt{1 + \frac{M}{100}} - 1$

Now:

Increase in the total cost = $\sqrt{\frac{2DC_1KI(1 - D/P)}{(K + I)}} - \sqrt{\frac{2DCKI(1 - D/P)}{(K + I)}}$

$$= \sqrt{\frac{2DCKI(1 - D/P)}{(K + I)}} \left[\sqrt{\frac{C_1}{C}} - 1 \right]$$

$$= \sqrt{\frac{2DCKI(1 - D/P)}{(K + I)}} \left[\sqrt{1 + \frac{M}{100}} - 1 \right]$$

And: % increase in $E = \sqrt{1 + \frac{M}{100}} - 1$

Similarly, increase in the maximum shortage quantity

$$= \sqrt{\frac{2DC_1I(1 - D/P)}{K(K + I)}} - \sqrt{\frac{2DCI(1 - D/P)}{K(K + I)}}$$

$$= \sqrt{\frac{2DCI(1 - D/P)}{K(K + I)}} \left[\sqrt{\frac{C_1}{C}} - 1 \right]$$

$$= \sqrt{\frac{2DCI(1 - D/P)}{K(K + I)}} \left[\sqrt{1 + \frac{M}{100}} - 1 \right]$$

And:

% increase in $J = \sqrt{1 + \frac{M}{100}} - 1$

Table 3.8 shows the summarised results.

Table 3.8: Results with respect to % setup cost increase including shortages.

Increase in the production batch size	$\sqrt{\frac{2DC(K + I)}{KI(1 - D/P)}} \left[\sqrt{1 + \frac{M}{100}} - 1 \right]$
% increase in the production batch size	$\sqrt{1 + \frac{M}{100}} - 1$
Increase in the total cost	$\sqrt{\frac{2DCKI(1 - D/P)}{(K + I)}} \left[\sqrt{1 + \frac{M}{100}} - 1 \right]$

Contd...

Contd...

% increase in the total cost	$\sqrt{1+\frac{M}{100}}-1$
Increase in the maximum shortage quantity	$\sqrt{\frac{2DCI(1-D/P)}{K(K+I)}}\left[\sqrt{1+\frac{M}{100}}-1\right]$
% increase in J	$\sqrt{1+\frac{M}{100}}-1$

3.5 Interaction of demand and setup cost

Because of demand or setup cost variation, lot size will not be similar. Production lot size increases due to demand increase as shown in Fig. 3.2. Higher batch size is represented by $Q1$ corresponding to an increased demand $D1$ because of a longer running time for facility related to production of an item. This is in comparison with reference demand D along with corresponding production batch size Q. If management wants to have similar lot size for operational ease, an option could be a reduction in setup cost if it is possible. In such case, a conscious effort may be made to reduce the setup cost.

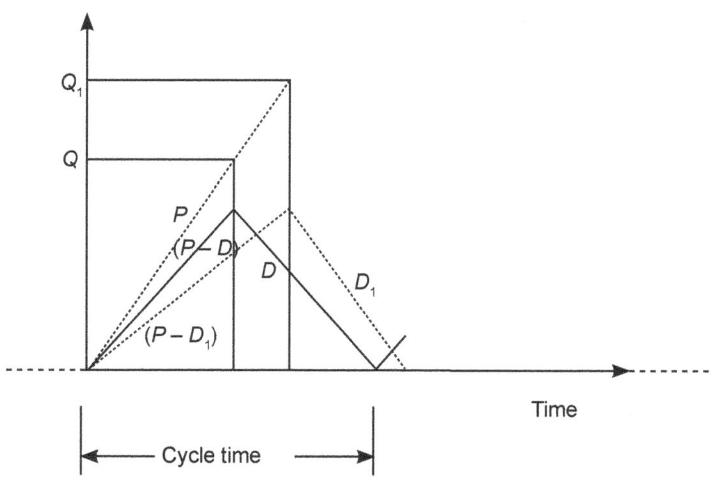

Figure 3.2: Higher batch size with demand increase

Example 3.6

With the following data:

Annual demand, $D = 600$ units

Setup cost, C = ₹ 45

Annual production rate, P = 960 units

Annual inventory carrying cost per unit, I = ₹ 40;

Production batch size:

$$Q = \sqrt{\frac{2DC}{(1 - D/P)I}}$$

$$= 60 \text{ units}$$

Now if demand is increased by 10%, i.e., D_1 = 660 units, then the optimal production batch size would be $Q1$ = 68.93

In order to keep similar batch size, the revised facility setup cost (C_1) can be obtained as:

$$\sqrt{\frac{2 \times 660 \times C_1}{(1 - 660/960) \times 40}} = 60$$

Or $$C_1 = ₹\ 34.09$$

That is approximately 24.24% decrease in setup cost. In order to generalise, let:

M = % increase in demand

N = % decrease in setup cost

And:

$$D_1 = D\left(1 + \frac{M}{100}\right)$$

$$C_1 = C\left(1 - \frac{N}{100}\right)$$

Now, for similar batch size:

$$\sqrt{\frac{2DC}{(1 - D/P)I}} = \sqrt{\frac{2D_1C_1}{(1 - D_1/P)I}}$$

Or $$\frac{DC}{(1 - D/P)} = \frac{D_1C_1}{(1 - D_1/P)}$$

Or $D_1C_1(1 - D/P) = DC(1 - C_1/P)$

Or $DC(1 + M/100)\ (1 - N/100)(1 - D/P) = DC[1 - (D/P)\ (1 + M/100)]$

Or $$1 - \frac{N}{100} = \frac{1 - (D/P)(1 + M/100)}{(1 + M/100)(1 - D/P)}$$

$$\text{Or } \frac{N}{100} = 1 - \frac{1-(D/P)(1+M/100)}{(1+M/100)(1-D/P)}$$

$$\text{Or } \frac{N}{100} = \frac{(1+M/100)(1-D/P) - [1-(D/P)(1+M/100)]}{(1+M/100)(1-D/P)}$$

$$\text{Or } \frac{N}{100} = \frac{(1+M/100) - (1+M/100)(D/P) - 1 + (D/P)(1+M/100)}{(1+M/100)(1-D/P)}$$

$$\text{Or } \frac{N}{100} = \frac{(M/100)}{(1+M/100)(1-D/P)}$$

$$\text{Or } N = \frac{M}{(1+M/100)(1-D/P)}$$

For the present case, values of N are provided in Table 3.9 corresponding to a given value of M. Values of N are higher than that of M, but slightly less sensitive toward higher values of M.

Table 3.9: Variation of N (setup cost) with respect to M (demand).

S. No.	M	$N = \frac{M}{(1+M/100)(1-D/P)}$
1	2	5.23
2	4	10.26
3	6	15.09
4	8	19.75
5	10	24.24

In the previous case, a change was initiated by demand parameter variation and the management response comprised of a setup cost reduction. However, if it is initiated by a setup cost reduction, then a suitable response could be in the form of a potential demand increase.

Now:

$$M = \text{\% reduction in setup cost}$$
$$N = \text{\% demand increase}$$

And:

$$C_1 = C\left(1 - \frac{M}{100}\right)$$

$$D_1 = D\left(1 + \frac{N}{100}\right)$$

Now, for similar batch size:

$$\sqrt{\frac{2DC}{(1-D/P)I}} = \sqrt{\frac{2D_1C_1}{(1-D_1/P)I}}$$

Or $\dfrac{DC}{(1-D/P)} = \dfrac{D_1C_1}{(1-D_1/P)}$

Or $D_1C_1(1-D/P) = DC(1-D_1/P)$

Or $DC(1-M/100)(1+N/100)(1-D/P) = DC[1-(D/P)(1+N/100)]$

Or $(1-M/100)(1+N/100)(1-D/P) = 1-(D/P)(1+N/100)$

Or $(1-N/100)(1+M/100)(1-D/P) = 1$

Or $1+\dfrac{N}{100} = \dfrac{1}{(1-M/100)-(D/P)(1-M/100)+(D/P)}$

Or $\dfrac{N}{100} = \dfrac{1}{(1-M/100)+(D/P)(M/100)} - 1$

Or $\dfrac{N}{100} = \dfrac{1-1+(M/100)-(D/P)(M/100)}{(1-M/100)+(D/P)(M/100)}$

Or $\dfrac{N}{100} = \dfrac{(M/100)(1-D/P)}{1-(M/100)(1-D/P)}$

Or $N = \dfrac{M(1-D/P)}{1-(M/100)(1-D/P)}$

For the present case, i.e, when change is triggered by setup cost reduction, values of N (response in the form of potential demand increase in order to maintain similar production batch size) are provided in Table 3.10 corresponding to a given value of M (% reduction in facility setup cost). Values of N are lower than that of M, but slightly more sensitive towards higher values of M.

Table 3.10: Variation of N (demand) with respect to M (setup cost).

S. No.	*M*	$N = \dfrac{M(1-D/P)}{1-(M/100)(1-D/P)}$
1	5	1.91
2	10	3.90
3	15	5.96
4	20	8.11
5	25	10.34

4
Inventory Holding Cost

Abstract : An attention is paid to the factors associated with the holding or carrying cost for the purpose of its estimation. Because of certain factors, an increase in the carrying cost is observed and it becomes necessary to determine the effects. After the development of necessary formulae, it has been observed that the percentage variation in total cost is higher than that in the batch size. Similarly the effects are determined with respect to a decrease in the carrying cost, and the percentage variation in batch size is observed to be more than that in the total cost. These have been proved analytically also.

Keyword : Inventory holding or carrying cost, factors for holding cost variation, percentage variation in output parameters, computational effects, analytical proof

After production of an item, inventory holding cost is borne by the organization until the item is sold or consumed.

4.1 Holding cost estimation

While estimating the cost of holding a product, attention should be paid to the factors as shown in Fig. 4.1.

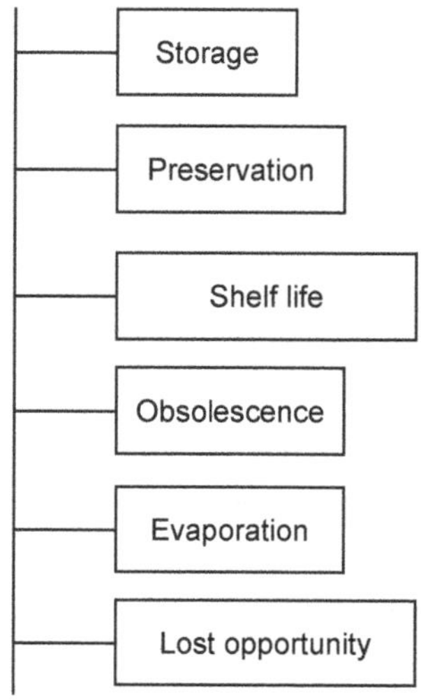

Figure 4.1: Factors associated with the holding cost

After production of an item, storage space is needed until the item is sold or consumed. If it is a finished product, it is sold in the market. If it is a component, it will eventually be consumed in the subsequent production process. That is, it will either be converted to the final product or will go to the finished product as a component. Therefore, certain waiting time is relevant in most of the cases. Because of this reason, storage cost is an important factor in the estimation of inventory holding cost.

Preservation of food items may be necessary and therefore such costs should be accounted in order to calculate the holding cost per unit item for a specified period. Such items and medicines are having certain shelf life, i.e., if these are held beyond that period, then those items expire. Refrigeration may also be necessary, and refrigeration cost/power charges need to be considered.

A component or subassembly might become obsolete if technology is changed in the context of engineering companies. Therefore, prediction of obsolescence also has a role to play in the issues concerning storage of inventories. In some of the products in the liquid state, or liquid products, evaporation is an issue that may also correspond to a rise in temperature. Evaporation losses need to be accounted that affects the holding cost of such items.

For example, because of space constraint (concerning the space occupied by one type of items), a manufacturing company could not produce another type of item. This is also a kind of lost opportunity. Similarly because of inventory carrying related to an item, the company should list other potential benefits that could not be obtained. Such analysis along with other factors will help in the holding cost estimation.

4.2 Increase in carrying cost

Increase in the carrying or holding cost may be because of:

(i) Higher storage cost
(ii) Higher preservation cost
(iii) Increased lost opportunity
(iv) Increased evaporation losses and also obsolescence

Example 4.1

With the following parameters:

Annual demand, $D = 600$ units

Setup cost, $C =$ ₹ 45

Annual production rate, $P = 960$ units

Annual inventory carrying cost per unit, I = ₹ 40;

Production batch size:

$$Q^* = \sqrt{\frac{2DC}{(1-D/P)I}}$$

$$= 60 \text{ units}$$

And total annual cost:

$$E^* = \sqrt{2DCI(1-D/P)}$$

$$= ₹\ 900$$

With the use of the base data as follows:

D	C	*I*	*P*	*Q*	*E*
600	45	40	960	60.00	900.00

Compute the effects of an increased holding cost as follows:

% Increase in I	5%	10%	15%	20%	25%	30%
I	42	44	46	48	50	52

The effects are shown in Table 4.1. Optimum production batch size reduces, however, total cost increases.

Table 4.1: Effects on parameters with respect to increase in carrying cost.

% Increase in I	5%	10%	15%	20%	25%	30%
I	42	44	46	48	50	52
Q	58.554	57.208	55.950	54.772	53.666	52.623
% Decrease in Q	2.41%	4.65%	6.75%	8.71%	10.56%	12.29%
E	922.23	943.93	965.14	985.90	1006.23	1026.16
% Increase in E	2.47%	4.88%	7.24%	9.54%	11.80%	14.02%

For a general approach, let:

M = % increase in annual inventory holding cost per unit (I)

And:

$$I_1 = \left(1+\frac{M}{100}\right)I$$

Reduction in production batch size = $\sqrt{\frac{2DC}{(1-D/P)I}} - \sqrt{\frac{2DC}{(1-D/P)I_1}}$

$$= \sqrt{\frac{2DC}{(1-D/P)I}}\left[1-\sqrt{\frac{I}{I_1}}\right]$$

$$= \sqrt{\frac{2DC}{(1-D/P)I}}\left[1-\sqrt{\frac{1}{(1+M/100)}}\right]$$

And:

$$\% \text{ reduction in } Q = 1-\sqrt{\frac{1}{(1+M/100)}}$$

Now:

$$\text{Increase in total cost} = \sqrt{2DCI_1(1-D/P)} - \sqrt{2DCI(1-D/P)}$$

$$= \sqrt{2DCI(1-D/P)}\left[\sqrt{\frac{I_1}{I}}-1\right]$$

$$= \sqrt{2DCI(1-D/P)}\left[\sqrt{(1+M/100)}-1\right]$$

And:

$$\% \text{ increase in } E = \sqrt{(1+M/100)}-1$$

The obtained generalised results are summarised in Table 4.2.

Table 4.2: Results with reference to % increase in carrying cost.

Reduction in the production batch size	$\sqrt{\frac{2DC}{(1-D/P)I}}\left[1-\sqrt{\frac{1}{(1+M/100)}}\right]$
% reduction in the production batch size	$1-\sqrt{\frac{1}{(1+M/100)}}$
Increase in total cost	$\sqrt{2DCI(1-D/P)}\left[\sqrt{(1+M/100)}-1\right]$
% increase in total cost	$\sqrt{(1+M/100)}-1$

Refer Table 4.1. % variation in total cost is higher than that in the batch size. Analytically also:

$$\sqrt{(1+M/100)}-1 > 1-\sqrt{\frac{1}{(1+M/100)}}$$

Or $$\sqrt{(1+M/100)}+\sqrt{\frac{1}{(1+M/100)}} > 2$$

Or $(1+M/100)+1>2\sqrt{(1+M/100)}$

Or $(1+M/100)+1>2\sqrt{(1+M/100)}$

Or $2+M/100>2\sqrt{(1+M/100)}$

Or $4+(M/100)^2+(M/25)>4+(M/25)$

Or $(M/100)^2>0$

And this is true.

4.3 Decrease in carrying cost

Decrease in the carrying or holding cost may be because of:

(i) Lower storage cost
(ii) Lower preservation cost
(iii) Decreased lost opportunity
(iv) Decreased evaporation losses and also obsolescence

Example 4.2

With the use of the following parameters:

D	*C*	*I*	*P*	*Q*	*E*
600	45	40	960	60.00	900.00

Compute the effects of a decreased holding cost as follows:

% Decrease in I	5%	10%	15%	20%	25%	30%
I	38	36	34	32	30	28

The effects are shown in Table 4.3. Optimum production batch size increases, however, total cost decreases.

Table 4.3: Effects on parameters with respect to decrease in carrying cost.

% Decrease in I	5%	10%	15%	20%	25%	30%
I	38	36	34	32	30	28
Q	61.559	63.246	65.079	67.082	69.282	71.714
% Increase in Q	2.60%	5.41%	8.47%	11.80%	15.47%	19.52%
E	877.21	853.81	829.76	804.98	779.42	752.99
% Decrease in E	2.53%	5.13%	7.80%	10.56%	13.40%	16.33%

For a general approach, let:

M = % decrease in annual inventory holding cost per unit (I)

And:

$$I_1 = \left(1 - \frac{M}{100}\right) I$$

Increase in production batch size = $\sqrt{\frac{2DC}{(1-D/P)I_1}} - \sqrt{\frac{2DC}{(1-D/P)I}}$

$$= \sqrt{\frac{2DC}{(1-D/P)I}}\left[\sqrt{\frac{I}{I_1}} - 1\right]$$

$$= \sqrt{\frac{2DC}{(1-D/P)I}}\left[\sqrt{\frac{1}{(1-M/100)}} - 1\right]$$

And:

% increase in $Q = \sqrt{\frac{1}{(1-M/100)}} - 1$

Now:

Reduction in total cost = $\sqrt{2DCI(1-D/P)} - \sqrt{2DCI_1(1-D/P)}$

$$= \sqrt{2DCI(1-D/P)}\left[1 - \sqrt{\frac{I_1}{I}}\right]$$

$$= \sqrt{2DCI(1-D/P)}\left[1 - \sqrt{(1-M/100)}\right]$$

And:

% reduction in $E = 1 - \sqrt{(1-M/100)}$

The obtained generalised results are also summarised in Table 4.4.

Table 4.4: Results with reference to % reduction in carrying cost.

Increase in the production batch size	$\sqrt{\frac{2DC}{(1-D/P)I}}\left[\sqrt{\frac{1}{(1-M/100)}} - 1\right]$
% increase in the production batch size	$\sqrt{\frac{1}{(1-M/100)}} - 1$
Reduction in total cost	$\sqrt{2DCI(1-D/P)}\left[1 - \sqrt{(1-M/100)}\right]$
% reduction in total cost	$1 - \sqrt{(1-M/100)}$

Refer Table 4.3. % variation in batch size is more than that in the total cost. Analytically also:

$$\sqrt{\frac{1}{(1-M/100)}}-1>1-\sqrt{(1-M/100)}$$

Or $$\sqrt{(1-M/100)}+\sqrt{\frac{1}{(1-M/100)}}>2$$

Or $$(1-M/100)+1>2\sqrt{(1-M/100)}$$

Or $$2-M/100>2\sqrt{(1-M/100)}$$

Or $$4+(M/100)^2-(M/25)>4(1-M/100)$$

Or $$4+(M/100)^2-(M/25)>4-(M/25)$$

Or $$(M/100)^2>0$$

And this is true.

4.4 Backlogging scenario

When all the shortages in the manufacturing system are backlogged, then the effects of holding cost variation are examined in this section.

4.4.1 Holding cost increase

Example 4.3

With the use of the following information:

P	*D*	*C*	*I*	*K*	*Q*	*E*	*J*
960	600	45	40	100	71.0	760.64	7.606388

Effects of holding cost are examined considering its variation as:

% Increase in I	5%	10%	15%	20%	25%	30%
I	42	44	46	48	50	52

Table 4.5 shows the corresponding variation in output parameters. Batch size decreases, whereas backlogged quantity and total cost increase. However, % variation in *E* and *J* is similar.

Table 4.5: Effects on parameters with shortages related to holding cost increase.

% Increase in I	5%	10%	15%	20%	25%	30%
I	42	44	46	48	50	52
Q	69.775	68.649	67.605	66.633	65.727	64.879
% Decrease in Q	1.73%	3.31%	4.78%	6.15%	7.43%	8.62%
E	773.91	786.61	798.76	810.41	821.58	832.32
% Increase in E	1.75%	3.41%	5.01%	6.54%	8.01%	9.42%
J	7.74	7.87	7.99	8.10	8.22	8.32
% Increase in J	1.75%	3.41%	5.01%	6.54%	8.01%	9.42%

In order to generalise:

$$I_1 = \left(1 + \frac{M}{100}\right) I$$

(i) Decrease in production batch size $= \sqrt{\frac{2DC(K+I)}{KI(1-D/P)}} - \sqrt{\frac{2DC(K+I_1)}{KI_1(1-D/P)}}$

$$= \sqrt{\frac{2DC(K+I)}{KI(1-D/P)}}\left[1 - \sqrt{\frac{I(K+I_1)}{I_1(K+I)}}\right]$$

$$= \sqrt{\frac{2DC(K+I)}{KI(1-D/P)}}\left[1 - \sqrt{\frac{K+I(1+M/100)}{(K+I)(1+M/100)}}\right]$$

And:

$$\%\text{ decrease in } Q = 1 - \sqrt{\frac{K+I(1+M/100)}{(K+I)(1+M/100)}}$$

(ii) Increase in the total cost $= \sqrt{\frac{2DCKI_1(1-D/P)}{(K+I_1)}} - \sqrt{\frac{2DCKI(1-D/P)}{(K+I)}}$

$$= \sqrt{\frac{2DCKI(1-D/P)}{(K+I)}}\left[\sqrt{\frac{I_1(K+I)}{I(K+I_1)}} - 1\right]$$

$$= \sqrt{\frac{2DCKI(1-D/P)}{(K+I)}}\left[\sqrt{\frac{(1+M/100)(K+I)}{K+I(1+M/100)}} - 1\right]$$

And:

$$\%\text{ increase in } E = \sqrt{\frac{(1+M/100)(K+I)}{K+I(1+M/100)}} - 1$$

(iii) Increase in backlogged quantity

$$= \sqrt{\frac{2DCI_1(1-D/P)}{K(K+I_1)}} - \sqrt{\frac{2DCI(1-D/P)}{K(K+I)}}$$

$$= \sqrt{\frac{2DCI(1-D/P)}{K(K+I)}}\left[\sqrt{\frac{I_1(K+I)}{I(K+I_1)}} - 1\right]$$

$$= \sqrt{\frac{2DCI(1-D/P)}{K(K+I)}}\left[\sqrt{\frac{(1+M/100)(K+I)}{K+I(1+M/100)}} - 1\right]$$

And:

$$\% \text{ increase in } J = \sqrt{\frac{(1+M/100)(K+I)}{K+I(1+M/100)}} - 1$$

Table 4.6 shows the summarised results.

Table 4.6: Results with respect to % holding cost increase including shortages.

Decrease in the production batch size	$\sqrt{\frac{2DC(K+I)}{KI(1-D/P)}}\left[1-\sqrt{\frac{K+I(1+M/100)}{(K+I)(1+M/100)}}\right]$
% decrease in the production batch size	$1-\sqrt{\frac{K+I(1+M/100)}{(K+I)(1+M/100)}}$
Increase in the total cost	$\sqrt{\frac{2DCKI(1-D/P)}{(K+I)}}\left[\sqrt{\frac{(1+M/100)(K+I)}{K+I(1+M/100)}}-1\right]$
% increase in the total cost	$\sqrt{\frac{(1+M/100)(K+I)}{K+I(1+M/100)}}-1$
Increase in the backlogged quantity	$\sqrt{\frac{2DCI(1-D/P)}{K(K+I)}}\left[\sqrt{\frac{(1+M/100)(K+I)}{K+I(1+M/100)}}-1\right]$
% increase in J	$\sqrt{\frac{(1+M/100)(K+I)}{K+I(1+M/100)}}-1$

4.4.2 Holding cost decrease

Example 4.4

With the use of the following information:

P	*D*	*C*	*I*	*K*	*Q*	*E*	*J*
960	600	45	40	100	71.0	760.64	7.606388

Effects of holding cost are examined considering its variation as:

% Decrease in I	5%	10%	15%	20%	25%	30%
I	38	36	34	32	30	28

Table 4.7 shows the corresponding variation in output parameters. Batch size increases, whereas backlogged quantity and total cost decrease. However, % variation in E and J is similar.

Table 4.7: Effects on parameters with shortages related to holding cost decrease.

% Decrease in I	5%	10%	15%	20%	25%	30%
I	38	36	34	32	30	28
Q	72.315	73.756	75.335	77.071	78.994	81.135
% Increase in Q	1.86%	3.89%	6.12%	8.56%	11.27%	14.29%
E	746.73	732.14	716.80	700.65	683.60	665.56
% Decrease in E	1.83%	3.75%	5.76%	7.89%	10.13%	12.50%
J	7.47	7.32	7.17	7.01	6.84	6.66
% Decrease in J	1.83%	3.75%	5.76%	7.89%	10.13%	12.50%

In order to generalise:

$$I_1 = \left(1 - \frac{M}{100}\right) I$$

(i) Increase in production batch size $= \sqrt{\frac{2DC(K+I_1)}{KI_1(1-D/P)}} - \sqrt{\frac{2DC(K+I)}{KI(1-D/P)}}$

$$= \sqrt{\frac{2DC(K+I)}{KI(1-D/P)}}\left[\sqrt{\frac{I(K+I_1)}{I_1(K+I)}} - 1\right]$$

$$= \sqrt{\frac{2DC(K+I)}{KI(1-D/P)}}\left[\sqrt{\frac{K+I(1-M/100)}{(K+I)(1-M/100)}} - 1\right]$$

And:

$$\%\text{ increase in } Q = \sqrt{\frac{K+I(1-M/100)}{(K+I)(1-M/100)}} - 1$$

(ii) Decrease in the total cost $= \sqrt{\dfrac{2DCKI(1-D/P)}{(K+I)}} - \sqrt{\dfrac{2DCKI_1(1-D/P)}{(K+I_1)}}$

$$= \sqrt{\frac{2DCKI(1-D/P)}{(K+I)}}\left[1-\sqrt{\frac{I_1(K+I)}{I(K+I_1)}}\right]$$

$$= \sqrt{\frac{2DCKI(1-D/P)}{(K+I)}}\left[1-\sqrt{\frac{(1-M/100)(K+I)}{K+I(1-M/100)}}\right]$$

And:

$$\% \text{ decrease in } E = 1-\sqrt{\frac{(1-M/100)(K+I)}{K+I(1-M/100)}}$$

(iii) Decrease in backlogged quantity

$$= \sqrt{\frac{2DCI(1-D/P)}{K(K+I)}} - \sqrt{\frac{2DCI_1(1-D/P)}{K(K+I_1)}}$$

$$= \sqrt{\frac{2DCI(1-D/P)}{K(K+I)}}\left[1-\sqrt{\frac{I_1(K+I)}{I(K+I_1)}}\right]$$

$$= \sqrt{\frac{2DCI(1-D/P)}{K(K+I)}}\left[1-\sqrt{\frac{(1-M/100)(K+I)}{K+I(1-M/100)}}\right]$$

And:

$$\% \text{ decrease in } J = 1-\sqrt{\frac{(1-M/100)(K+I)}{K+I(1-M/100)}}$$

Table 4.8 shows the summarised results.

Table 4.8: Results with respect to % holding cost decrease including shortages.

Increase in the production batch size	$\sqrt{\dfrac{2DC(K+I)}{KI(1-D/P)}}\left[\sqrt{\dfrac{K+I(1-M/100)}{(K+I)(1-M/100)}}-1\right]$
% increase in the production batch size	$\sqrt{\dfrac{K+I(1-M/100)}{(K+I)(1-M/100)}}-1$
Decrease in the total cost	$\sqrt{\dfrac{2DCKI(1-D/P)}{(K+I)}}\left[1-\sqrt{\dfrac{(1-M/100)(K+I)}{K+I(1-M/100)}}\right]$

% decrease in the total cost	$1-\sqrt{\frac{(1-M/100)(K+I)}{K+I(1-M/100)}}$
Decrease in the backlogged quantity	$\sqrt{\frac{2DCI(1-D/P)}{K(K+I)}}\left[1-\sqrt{\frac{(1-M/100)(K+I)}{K+I(1-M/100)}}\right]$
% decrease in J	$1-\sqrt{\frac{(1-M/100)(K+I)}{K+I(1-M/100)}}$

4.5 Interaction of carrying cost with other parameters

In order to examine the interaction of parameters, examples and analytical results are provided.

4.5.1 Without backlogging

When shortages are not allowed and a change in the production system is initiated by a variation in holding cost, then the management response belongs to the variation in other parameters for the objective of:

(i) Similar production batch size

(ii) Similar total cost

4.5.1.1 Change initiated by holding cost

This is explained with the help of examples.

(a) Holding cost increase

Example 4.5

An available set of information is as follows:

Annual demand, $D = 600$ units

Setup cost, $C =$ ₹ 45

Annual production rate, $P = 960$ units

Annual inventory carrying cost per unit, $I =$ ₹ 40;

Production batch size:

$$Q^* = \sqrt{\frac{2DC}{(1-D/P)I}}$$

$$= 60 \text{ units}$$

And total annual cost:

$$E^* = \sqrt{2DCI(1 - D/P)}$$
$$= ₹\,900$$

Now if inventory holding cost is increased by 10%, then:

$$I_1 = ₹\,44$$

Revised output parameters are as follows:

$$Q = 57.21$$
$$E = ₹\,943.93$$

In order to have similar lot size:

$$\sqrt{\frac{2 \times 600 \times 45}{(1 - 600/960) \times 40}} = \sqrt{\frac{2 \times D_1 \times 45}{(1 - D_1/960) \times 44}}$$

Or $D_1 = 621.18$ units

In case where objective is to keep similar total cost, then the ordering cost may be reduced along with the holding cost increase, i.e.,:

$$900 = \sqrt{2 \times 600 \times C_1 \times 44(1 - 600/960)}$$

Or $C_1 = ₹\,40.91$

For a general approach, let:

M = % increase in holding cost

N = % increase in demand

Therefore,

$$I_1 = \left(1 + \frac{M}{100}\right) I$$

$$D_1 = \left(1 + \frac{N}{100}\right) D$$

For similar lot size:

$$\sqrt{\frac{2DC}{(1 - D/P)I}} = \sqrt{\frac{2D_1C}{(1 - D_1/P)I_1}}$$

Or $$\frac{D}{(1 - D/P)I} = \frac{D_1}{(1 - D_1/P)I_1}$$

Or $D_1I(1 - D/P) = DI_1(1 - D_1/P)$

Or $D_1I - D_1ID/P - DI_1 + DI_1D_1/P = 0$

Or $D_1[I - ID/P + DI_1/P] = DI_1$

Or $$D_1 = \frac{DI_1}{I - ID/P + DI_1/P}$$

Or $\quad 1+\dfrac{N}{100}=\dfrac{I_1}{I-ID/P+DI_1/P}$

Or $\quad \dfrac{N}{100}=\dfrac{I_1-I+ID/P-DI_1/P}{I-ID/P+DI_1/P}$

Or $\quad \dfrac{N}{100}=\dfrac{(1+M/100)I-I+ID/P-(1+M/100)ID/P}{I-ID/P+(1+M/100)ID/P}$

Or $\quad \dfrac{N}{100}=\dfrac{(MI/100)-(MID/100P)}{I+(MID/100P)}$

Or $\quad \dfrac{N}{100}=\dfrac{(MI/100)(1-D/P)}{I(1+MD/100P)}$

Or $\quad N=\dfrac{M(1-D/P)}{1+(MD/100P)}$

For the present case, values of N are provided in Table 4.9 corresponding to a given value of M. Values of N are lower than that of M, and also less sensitive towards higher values of M.

Table 4.9: Variation of N (demand) with respect to M (holding cost).

S. No.	*M*	$N=\dfrac{M(1-D/P)}{1+(MD/100P)}$
1	10	3.53
2	15	5.14
3	20	6.67
4	25	8.11
5	30	9.47

When the objective is similar total cost, a response might be setup cost reduction.

M = % increase in holding cost

N = % reduction in setup cost

Therefore,

$$I_1 = \left(1+\frac{M}{100}\right)I$$

$$C_1 = \left(1-\frac{N}{100}\right)C$$

For similar total cost:

$$\sqrt{2DCI(1-D/P)} = \sqrt{2DC_1I_1(1-D/P)}$$

Or
$$CI = C_1I_1$$

Or
$$C_1 = \frac{CI}{I_1}$$

Or
$$1-\frac{N}{100} = \frac{1}{(1+M/100)}$$

Or
$$\frac{N}{100} = 1-\frac{1}{(1+M/100)}$$

Or
$$\frac{N}{100} = \frac{(M/100)}{(1+M/100)}$$

Or
$$N = \frac{M}{(1+M/100)}$$

For the present case, values of N are provided in Table 4.10 corresponding to a given value of M. Values of N are lower than that of M and also less sensitive toward higher values of M.

Table 4.10: Variation of N (setup cost) with respect to M (holding cost).

S. No.	M	$N = \frac{M}{(1+M/100)}$
1	10	9.09
2	15	13.04
3	20	16.67
4	25	20
5	30	23.08

(b) Holding cost decrease

Setup cost reduction can be a management response with the objective of similar batch size.

$$I_1 = \left(1-\frac{M}{100}\right)I$$

$$C_1 = \left(1-\frac{N}{100}\right)C$$

Now:

$$\sqrt{\frac{2DC}{(1-D/P)I}} = \sqrt{\frac{2DC_1}{(1-D/P)I_1}}$$

Or $$C_1 = \frac{CI_1}{I}$$

Or $$1-\frac{N}{100} = 1-\frac{M}{100}$$

Or $$N = M$$

% variation in both parameters is similar in this case.

4.5.1.2 Change initiated by other parameters

Demand and setup cost and their variations have been studied in previous chapters.

(a) Change initiated by demand

When demand reduces, a suitable response is the holding cost decrease for a similar batch size.

$$D_1 = D\left(1-\frac{M}{100}\right)$$

$$I_1 = I\left(1-\frac{N}{100}\right)$$

$$\sqrt{\frac{2DC}{(1-D/P)I}} = \sqrt{\frac{2D_1C}{(1-D_1/P)I_1}}$$

Or $$\frac{D}{(1-D/P)I} = \frac{D_1}{(1-D_1/P)I_1}$$

Or $$I_1D(1-D_1/P) = ID_1(1-D/P)$$

Or $$I_1 = \frac{ID_1(1-D/P)}{D(1-D_1/P)}$$

Or $$1-\frac{N}{100} = \frac{D_1(1-D/P)}{D(1-D_1/P)}$$

Or $$1-\frac{N}{100} = \frac{(1-M/100)(1-D/P)}{1-(1-M/100)(D/P)}$$

Or $$\frac{N}{100} = 1-\frac{(1-M/100)(1-D/P)}{(1-D/P)+(M/100)(D/P)}$$

Or $$\frac{N}{100} = \frac{(1-D/P)+(M/100)(D/P)-(1-M/100)(1-D/P)}{(1-D/P)+(M/100)(D/P)}$$

Or $$\frac{N}{100} = \frac{(M/100)(D/P)+(M/100)(1-D/P)}{(1-D/P)+(M/100)(D/P)}$$

Or $$\frac{N}{100} = \frac{(M/100)}{(1-D/P)+(M/100)(D/P)}$$

Or $$N = \frac{M}{(1-D/P)+(M/100)(D/P)}$$

For the present case, i.e., when:

Annual demand, $D = 600$ units

Annual production rate, $P = 960$ units

Values of N are provided in Table 4.11, corresponding to a given value of M. Values of N are higher than that of M and less sensitive toward higher values of M.

Table 4.11: Variation of N (holding cost) with respect to M (demand).

S. No.	M	$N = \frac{M}{(1-D/P)+(M/100)(D/P)}$
1	2	5.16
2	4	10.00
3	6	14.55
4	8	18.82
5	10	22.86

(a) Change initiated by setup cost

For similar total cost, a holding cost reduction is a response when setup cost is increased.

$$C_1 = C_1 = C\left(1+\frac{M}{100}\right)$$

$$I_1 = I\left(1-\frac{N}{100}\right)$$

$$\sqrt{2DCI(1-D/P)} = \sqrt{2DC_1I_1(1-D/P)}$$

Or $$I_1 = \frac{CI}{C_1}$$

Or $$1-\frac{N}{100} = \frac{C}{C_1}$$

Or $$1-\frac{N}{100} = \frac{1}{(1+M/100)}$$

Or $$\frac{N}{100} = 1-\frac{1}{(1+M/100)}$$

Or $$\frac{N}{100} = \frac{(M/100)}{(1+M/100)}$$

Or $$N = \frac{M}{(1+M/100)}$$

Table 4.12 shows the values of N that are corresponding to a given value of M. Values of N are lower than that of M and also less sensitive toward higher values of M.

Table 4.12: Variation of N (holding cost) with respect to M (setup cost).

S. No.	M	$N = \frac{M}{(1+M/100)}$
1	10	9.09
2	15	13.04
3	20	16.67
4	25	20.00
5	30	23.08

4.5.2 With backlogging

When all the shortages are fully backlogged, then the analytical results are presented in this section.

4.5.2.1 Change initiated by carrying cost

(a) Holding cost increase

Example 4.6

With the reference set of data as follows:

P	D	C	I	K	Q	E	J
960	600	45	40	100	71.0	760.64	7.606388

Increase the carrying cost by 10% to get the revised batch size as:

68.65 units

In order to have similar batch size, i.e., 71 units, the potentially increased demand is:

615 units

For a generalization:

$$I_1 = I\left(1+\frac{M}{100}\right)$$

$$D_1 = D\left(1+\frac{N}{100}\right)$$

For similar batch size:

$$\sqrt{\frac{2DC(K+I)}{KI(1-D/P)}} = \sqrt{\frac{2D_1C(K+I_1)}{KI_1(1-D_1/P)}}$$

Or $$\frac{D(K+I)}{I(1-D/P)} = \frac{D_1(K+I_1)}{I_1(1-D_1/P)}$$

Or $$D_1I(K+I_1)(1-D/P) = DI_1(K+I)(1-D_1/P)$$

Or $$D_1I(K+I_1)(1-D/P) = DI_1(K+I) - D_1I_1(K+I)(D/P)$$

Or $$D_1\left[I(K+I_1)(1-D/P) + I_1(K+I)(D/P)\right] = DI_1(K+I)$$

Or $$D_1 = \frac{DI_1(K+I)}{I(K+I_1)(1-D/P)+I_1(K+I)(D/P)}$$

Or $$1+\frac{N}{100} = \frac{I_1(K+I)}{I(K+I_1)(1-D/P)+I_1(K+I)(D/P)}$$

Or $$\frac{N}{100} = \frac{I_1(K+I) - I(K+I_1)(1-D/P) - I_1(K+I)(D/P)}{I(K+I_1)(1-D/P)+I_1(K+I)(D/P)}$$

Or $$\frac{N}{100} = \frac{(1+M/100)(K+I) - (K+I_1)(1-D/P) - (1+M/100)(K+I)(D/P)}{(K+I_1)(1-D/P)+(1+M/100)(K+I)(D/P)}$$

Or $$\frac{N}{100} = \frac{(1+M/100)(K+I)(1-D/P) - \left[K+I(1+M/100)\right](1-D/P)}{\left[K+I(1+M/100)\right](1-D/P)+(1+M/100)(K+I)(D/P)}$$

Or $$\frac{N}{100} = \frac{(1-D/P)\left[(1+M/100)(K+I) - K - I(1+M/100)\right]}{K(1-D/P)+I(1+M/100)(1-D/P)+K(1+M/100)(D/P)+I(1+M/100)(D/P)}$$

Or $$\frac{N}{100}=\frac{(1-D/P)\left[K(1+M/100)-K\right]}{K(1-D/P)+K(1+M/100)(D/P)+I(1+M/100)\left[1-(D/P)+(D/P)\right]}$$

Or $$\frac{N}{100}=\frac{(1-D/P)K\left[(1+M/100)-1\right]}{K-(KD/P)+(KD/P)+(KD/P)(M/100)+I(1+M/100)}$$

Or $$\frac{N}{100}=\frac{(1-D/P)(KM/100)}{K\left[1+(D/P)(M/100)\right]+I(1+M/100)}$$

Or $$N=\frac{KM(1-D/P)}{K+I+(M/100)\left[(KD/P)+I\right]}$$

Table 4.13 shows the values of N that are corresponding to a given value of M related to the present example. Values of N are much lower than that of M and also less sensitive toward higher values of M.

Table 4.13: Variation of N (demand) corresponding to M (carrying cost) with backlog.

S. No.	M	$N=\frac{KM(1-D/P)}{K+I+(M/100)\left[(KD/P)+I\right]}$
1	10	2.50
2	15	3.62
3	20	4.67
4	25	5.66
5	30	6.59

When objective is similar total cost, setup cost reduction might be a response. Now:

$$I_1=I\left(1+\frac{M}{100}\right)$$

$$C_1=C\left(1-\frac{N}{100}\right)$$

$$\sqrt{\frac{2DCKI(1-D/P)}{(K+I)}}=\sqrt{\frac{2DC_1KI_1(1-D/P)}{(K+I_1)}}$$

Or $$\frac{CI}{(K+I)}=\frac{C_1I_1}{(K+I_1)}$$

Or $$C_1 = \frac{(K+I_1)CI}{(K+I)I_1}$$

Or $$1 - \frac{N}{100} = \frac{(K+I_1)I}{(K+I)I_1}$$

Or $$1 - \frac{N}{100} = \frac{K+I(1+M/100)}{(K+I)(1+M/100)}$$

Or $$\frac{N}{100} = \frac{(K+I)(1+M/100) - K - I(1+M/100)}{(K+I)(1+M/100)}$$

Or $$\frac{N}{100} = \frac{K(1+M/100) - K}{(K+I)(1+M/100)}$$

Or $$\frac{N}{100} = \frac{(KM/100)}{(K+I)(1+M/100)}$$

Or $$N = \frac{KM}{(K+I)(1+M/100)}$$

Table 4.14 shows the values of N that are corresponding to a given value of M related to the present situation. Values of N are lower than that of M and also less sensitive toward higher values of M.

Table 4.14: Variation of N (setup cost) corresponding to M (increased I) with backlog.

S. No.	M	$N = \frac{KM}{(K+I)(1+M/100)}$
1	10	6.49
2	15	9.32
3	20	11.90
4	25	14.29
5	30	16.48

(b) Holding cost decrease

Setup cost reduction can be a management response for similar batch size.

$$I_1 = \left(1 - \frac{M}{100}\right)I$$

$$C_1 = \left(1 - \frac{N}{100}\right)C$$

$$\sqrt{\frac{2DC(K+I)}{KI(1-D/P)}} = \sqrt{\frac{2DC_1(K+I_1)}{KI_1(1-D/P)}}$$

Or $$\frac{C(K+I)}{I} = \frac{C_1(K+I_1)}{I_1}$$

Or $$C_1 = \frac{I_1 C(K+I)}{I(K+I_1)}$$

Or $$1-\frac{N}{100} = \frac{I_1(K+I)}{I(K+I_1)}$$

Or $$1-\frac{N}{100} = \frac{(1-M/100)(K+I)}{K+I(1-M/100)}$$

Or $$\frac{N}{100} = \frac{K+I(1-M/100)-(1-M/100)(K+I)}{K+I(1-M/100)}$$

Or $$\frac{N}{100} = \frac{K-K(1-M/100)}{K+I(1-M/100)}$$

Or $$N = \frac{KM}{K+I(1-M/100)}$$

With the reference set of data and considering the relevant values of K and I as 100 and 40 respectively, values of N are shown in Table 4.15. Values of N are lower than that of M but are more sensitive towards higher values of M.

Table 4.15: Variation of N (setup cost) corresponding to M (decreased I) with backlog.

S. No.	M	$N = \frac{KM}{K+I(1-M/100)}$
1	10	7.35
2	15	11.19
3	20	15.15
4	25	19.23
5	30	23.44

4.5.2.2 Change initiated by other parameters

(a) Change initiated by demand

With demand reduction, holding cost decrease is a possible response for similar production batch size.

$$D_1 = D\left(1-\frac{M}{100}\right)$$

$$I_1 = I\left(1-\frac{N}{100}\right)$$

$$\sqrt{\frac{2DC(K+I)}{KI(1-D/P)}} = \sqrt{\frac{2D_1C(K+I_1)}{KI_1(1-D_1/P)}}$$

Or $$\frac{D(K+I)}{I(1-D/P)} = \frac{D_1(K+I_1)}{I_1(1-D_1/P)}$$

Or $$DI_1(K+I)(1-D_1/P) = D_1I(K+I_1)(1-D/P)$$

Or $$DI_1(K+I)(1-D_1/P) = D_1IK(1-D/P) + D_1II_1(1-D/P)$$

Or $$I_1\left[D(K+I)(1-D_1/P) - D_1I(1-D/P)\right] = D_1IK(1-D/P)$$

Or $$I_1 = \frac{D_1IK(1-D/P)}{D(K+I)(1-D_1/P) - D_1I(1-D/P)}$$

Or $$1-\frac{N}{100} = \frac{D_1K(1-D/P)}{D(K+I)(1-D_1/P) - D_1I(1-D/P)}$$

Or $$1-\frac{N}{100} = \frac{K(1-M/100)(1-D/P)}{(K+I)(1-D_1/P) - I(1-M/100)(1-D/P)}$$

Or $$\frac{N}{100} = \frac{(K+I)(1-D_1/P) - I(1-M/100)(1-D/P) - K(1-M/100)(1-D/P)}{(K+I)(1-D_1/P) - I(1-M/100)(1-D/P)}$$

Or $$\frac{N}{100} = \frac{(K+I)(1-D_1/P) - (K+I)(1-M/100)(1-D/P)}{K(1-D_1/P) + I(1-D_1/P) - I(1-M/100)(1-D/P)}$$

Or $$\frac{N}{100} = \frac{(K+I)\left[1-(1-M/100)(D/P) - (1-M/100)(1-D/P)\right]}{K\left[1-(1-M/100)(D/P)\right] + I\left[1-(1-M/100)(D/P) - (1-M/100)(1-D/P)\right]}$$

Or $$\frac{N}{100} = \frac{(K+I)\left[1-(1-M/100)\right]}{K\left[1-(1-M/100)(D/P)\right] + I\left[1-(1-M/100)\{(D/P)+(1-D/P)\}\right]}$$

Or $$\frac{N}{100} = \frac{(K+I)(M/100)}{K(1-D/P) + (KD/P)(M/100) + I\left[1-(1-M/100)\right]}$$

Or $$N = \frac{M(K+I)}{K(1-D/P)+(KD/P)(M/100)+I(M/100)}$$

Or $$N = \frac{M(K+I)}{K(1-D/P)+(I+KD/P)(M/100)}$$

With the reference set of data as follows:

P	*D*	*C*	*I*	*K*	*Q*	*E*	*J*
960	600	45	40	100	71.0	760.64	7.606388

The values of *N* are shown in Table 4.16. Values of *N* are much higher than that of *M* but are less sensitive toward larger values of M.

Table 4.16: Variation of *N* (holding cost) corresponding to *M* (decreased *D*) with backlog.

S. No.	*M*	$N = \frac{M(K+I)}{K(1-D/P)+(I+KD/P)(M/100)}$
1	2	7.08
2	4	13.46
3	6	19.24
4	8	24.51
5	10	29.32

(b) Change initiated by setup cost

With setup cost reduction, a response is holding cost reduction for similar production batch size.

$$C_1 = C\left(1-\frac{M}{100}\right)$$

$$I_1 = I\left(1-\frac{N}{100}\right)$$

$$\sqrt{\frac{2DC(K+I)}{KI(1-D/P)}} = \sqrt{\frac{2DC_1(K+I_1)}{KI_1(1-D/P)}}$$

Or $$\frac{C(K+I)}{I} = \frac{C_1(K+I_1)}{I_1}$$

Or $$I_1C(K+I) - I_1C_1I = C_1KI$$

Or $$I_1 = \frac{C_1KI}{C(K+I)-C_1I}$$

Or $$1-\frac{N}{100}=\frac{K(1-M/100)}{(K+I)-I(1-M/100)}$$

Or $$\frac{N}{100}=\frac{(K+I)-I(1-M/100)-K(1-M/100)}{(K+I)-I(1-M/100)}$$

Or $$\frac{N}{100}=\frac{(K+I)-(K+I)(1-M/100)}{K+I-I+(IM/100)}$$

Or $$\frac{N}{100}=\frac{(K+I)(1-1+M/100)}{K+(IM/100)}$$

Or $$\frac{N}{100}=\frac{(K+I)(M/100)}{K+(IM/100)}$$

Or $$N=\frac{M(K+I)}{K+(IM/100)}$$

With reference to the present example, values of N are showin in Table 4.17. Values of N are higher than that of M but are less sensitive toward larger values of M.

Table 4.17: Variation of N (holding cost) corresponding to M (reduced C) with backlog.

S. No.	M	$N=\frac{M(K+I)}{K+(IM/100)}$
1	10	13.46
2	15	19.81
3	20	25.93
4	25	31.82
5	30	37.50

In case where objective is similar total cost and a change is initiated by setup cost increase, a suitable response can be in the form of holding cost reduction.

Now:

$$C_1 = C\left(1+\frac{M}{100}\right)$$

$$I_1 = I\left(1-\frac{N}{100}\right)$$

For similar total cost:

$$\sqrt{\frac{2DCKI(1-D/P)}{(K+I)}} = \sqrt{\frac{2DC_1KI_1(1-D/P)}{(K+I_1)}}$$

Or $$\frac{CI}{(K+I)} = \frac{C_1I_1}{(K+I_1)}$$

Or $$C_1I_1(K+I) = CKI + CII_1$$

Or $$I_1\left[C_1(K+I) - CI\right] = CKI$$

Or $$I_1 = \frac{CKI}{C_1(K+I) - CI}$$

Or $$1 - \frac{N}{100} = \frac{K}{(1+M/100)(K+I) - I}$$

Or $$\frac{N}{100} = \frac{(1+M/100)(K+I) - I - K}{(1+M/100)(K+I) - I}$$

Or $$\frac{N}{100} = \frac{(K+I)(1+M/100-1)}{K(1+M/100) + I(1+M/100) - I}$$

Or $$\frac{N}{100} = \frac{(K+I)(M/100)}{K(1+M/100) + (IM/100)}$$

Or $$N = \frac{M(K+I)}{K + (K+I)(M/100)}$$

With reference to the present case, values of N are shown in Table 4.18. Values of N are more than that of M but are less sensitive toward higher values of M.

Table 4.18: Variation of N (holding cost) corresponding to M (increased C) with backlog.

S. No.	M	$N = \frac{M(K+I)}{K+(K+I)(M/100)}$
1	10	12.28
2	15	17.36
3	20	21.87
4	25	25.93
5	30	29.58

5

Production Cost

Abstract : After a discussion related to the factors concerning the production cost, optimal values of the batch quantity and total annual cost have been derived. With an illustration, those data are further used for unit production cost variation, i.e., on lower and higher side. This is explained along with the reasons for a lower as well as a higher production cost. Percentage variation in production batch size is lower in case of a higher production cost as compared to that in case of a lower production cost. With stock out inclusion also, the generalised results have been obtained.

Keyword : Production cost, unit production cost variation, illustration, production batch size, generalised results

Various factors concerning the production cost per product are shown in Fig. 5.1.

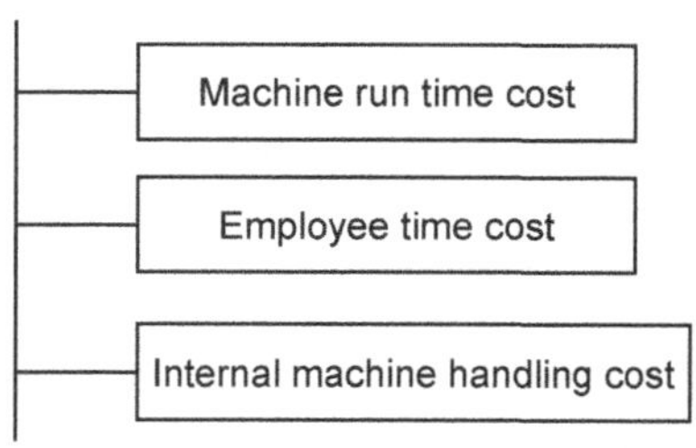

Figure 5.1: Factors concerning the production cost

In order to estimate the actual production cost per unit item, expenditure involved in running the machine after the facility setup should be accounted. This also includes the power charges. Employees time related to the production can be converted in terms of cost involved. In most of the cases, certain internal material handling efforts and related costs may suitably be attributed towards the production cost also.

If production cost is to be included explicitly in the determination of batch size and total cost, then the formulation needs to have some changes. While estimating the annual inventory carrying cost per unit (I), production cost per unit should be specifically visible. In certain cases, value added production cost per unit can also be incorporated. Now carrying cost (I) may be replaced by a multiplication of unit production cost and annual inventory carrying cost fraction. Annual production cost should also be added in the total relevant cost.

In order to generalise, let:

U = Unit production cost

F = Annual inventory carrying cost fraction

As the annual demand is D, the annual manufacturing or production cost:

$$AMC = UD \tag{5.1}$$

Annual inventory carrying cost is given by Eq. (1.3) as:

$$AIC = \frac{Q(1-D/P)I}{2}$$

Since I may be replaced by (UF),

$$AIC = \frac{Q(1-D/P)UF}{2} \tag{5.2}$$

Annual production setup cost is given by Eq. (1.4) as:

$$APC = \frac{DC}{Q}$$

Adding all these cost components, the total relevant cost:

$$E = AMC + AIC + APC$$

$$E = UD + \frac{Q(1-D/P)UF}{2} + \frac{DC}{Q} \tag{5.3}$$

In order to obtain the optimal production batch size, differentiate w.r.t. Q and equate to zero.

$$\frac{(1-D/P)UF}{2} - \frac{DC}{Q^2} = 0$$

Or
$$Q^2 = \frac{2DC}{(1-D/P)UF}$$

Therefore, the economic batch quantity:

$$Q^* = \sqrt{\frac{2DC}{(1-D/P)UF}} \tag{5.4}$$

Substituting in Eq. (5.3), the optimal total annual cost:

$$E^* = UD + \sqrt{2DCUF(1-D/P)} \tag{5.5}$$

Example 5.1

Consider the following data:

Annual demand, D = 600 units

Setup cost, C = ₹ 45

Annual production rate, P = 960 units

Unit production cost, U = ₹ 170

Annual inventory carrying cost fraction, F = 0.2

Now, from Eq. (5.4):

Optimal production batch size, Q = 65.08

And, from Eq. (5.5):

Optimal total annual cost, E = ₹ 102829.76

These data may further be used for unit production cost variation, i.e., on lower or higher side.

5.1 Lower production cost

Production cost might be lower because of the following reasons:

(i) In some cases, fuel is directly or indirectly used in running the machine. When fuel prices are reduced, the unit production cost is expected to be lower.

(ii) Because of retirement or transfer, a relatively junior employee might be deputed on certain facility. Therefore, production cost may be reduced as the salary/wages are comparatively lower.

(iii) As the low cost automation or an improved material handling practice is adopted, internal material handling cost related to production can be made lower. This results into a decreased overall unit production cost.

Example 5.2

Consider the input and output parameters of Example 5.1 as follows:

D	*C*	*P*	*U*	*F*	*Q*	*E*
600	45	960	170	0.2	65.08	102829.76

The unit production cost is lowered as follows:

% decrease in U	5%	10%	15%	20%	25%	30%
U	161.5	153	144.5	136	127.5	119

The effects on production batch size and total cost are shown in Table 5.1. A lower production cost affects the output parameters as:

(a) Increase in the production batch size

(b) Reduction in the total cost

Table 5.1: Effects on output parameters corresponding to production cost decrease.

% decrease in U	5%	10%	15%	20%	25%	30%
U	161.5	153	144.5	136	127.5	119
Q	66.77	68.60	70.59	72.76	75.15	77.78
% increase in Q	2.60%	5.41%	8.47%	11.80%	15.47%	19.52%
E	97708.75	92587.18	87465	82342.16	77218.59	72094.23
% decrease in E	4.98%	9.96%	14.94%	19.92%	24.91%	29.89%

In order to generalise, let

M = % reduction in unit production cost

$$U_1 = \left(1 - \frac{M}{100}\right)U$$

(i) Increase in production batch size

$$\sqrt{\frac{2DC}{(1-D/P)U_1F}} - \sqrt{\frac{2DC}{(1-D/P)UF}}$$

$$= \sqrt{\frac{2DC}{(1-D/P)UF}}\left[\sqrt{\frac{U}{U_1}} - 1\right]$$

$$= \sqrt{\frac{2DC}{(1-D/P)UF}}\left[\sqrt{\frac{1}{(1-M/100)}} - 1\right]$$

And:

$$\text{\% increase in } Q = \sqrt{\frac{1}{(1-M/100)}} - 1$$

(ii) Reduction in the total cost:

$$= UD + \sqrt{2DCUF(1-D/P)} - U_1D - \sqrt{2DCU_1F(1-D/P)}$$

$$= \left\{UD + \sqrt{2DCUF(1-D/P)}\right\}\left[1 - \frac{U_1D + \sqrt{2DCU_1F(1-D/P)}}{UD + \sqrt{2DCUF(1-D/P)}}\right]$$

$$= \left\{UD + \sqrt{2DCUF(1-D/P)}\right\}\left[\frac{D(U-U_1) + \sqrt{2DCF(1-D/P)}\left\{\sqrt{U} - \sqrt{U_1}\right\}}{UD + \sqrt{2DCUF(1-D/P)}}\right]$$

$$= \frac{DUM}{100} + \sqrt{2DCUF(1 - D/P)}\left\{1 - \sqrt{1 - \frac{M}{100}}\right\}$$

And:

$$\% \text{ reduction in } E = \frac{(DUM/100) + \sqrt{2DCUF(1-D/P)}\left\{1 - \sqrt{(1 - M/100)}\right\}}{UD + \sqrt{2DCUF(1-D/P)}}$$

Table 5.2 represents the generalised results.

Table 5.2: Results corresponding to % reduction in unit production cost.

Increase in production batch size	$\sqrt{\frac{2DC}{(1-D/P)UF}}\left[\sqrt{\frac{1}{(1-M/100)}} - 1\right]$
% Increase in Q	$\sqrt{\frac{1}{(1-M/100)}} - 1$
Reduction in total cost	$\frac{DUM}{100} + \sqrt{2DCUF(1-D/P)}\left\{1 - \sqrt{1 - \frac{M}{100}}\right\}$
% reduction in E	$\frac{(DUM/100) + \sqrt{2DCUF(1-D/P)}\left\{1 - \sqrt{(1-M/100)}\right\}}{UD + \sqrt{2DCUF(1-D/P)}}$

5.2 Higher production cost

There may be an increase in unit production cost because of the following reasons:

(i) Power charges may increase and the cost of running a machine for a specified period increases.

(ii) Because of the seniority of a worker employed on a machine, salary/ wages might be higher. This results into a higher production cost per unit item.

(iii) Internal material handling cost may depend on the device being used or the number of workers employed. The cost increases because of salary/ wages and/ or price of fuel/power.

Example 5.3

Consider the input and output parameters of Example 5.1 as follows:

D	*C*	*P*	*U*	*F*	*Q*	*E*
600	45	960	170	0.2	65.08	102829.76

The unit production cost is increased as given below:

% increase in U	5%	10%	15%	20%	20%	30%
U	178.5	187	195.5	204	212.5	221

The effect is examined and shown in Table 5.3. A lower production cost affects the output parameter as:

(a) Decrease in the production batch size

(b) Increase in the total cost

Table 5.3: Effects on output parameters corresponding to production cost increase.

% increase in U	5%	10%	15%	20%	25%	30%
U	178.5	187	195.5	204	212.5	221
Q	63.51	62.05	60.69	59.41	58.21	57.08
% decrease in Q	2.41%	4.65%	6.75%	8.71%	10.56%	12.29%
E	107950.25	113070.26	118189.822	123309.96	128427.70	133546.07
% increase in E	4.98%	9.96%	14.94%	19.92%	24.89%	29.87%

In order to generalise, let:

M = % increase in unit production cost

$$U_1 = \left(1+\frac{M}{100}\right)U$$

(i) Reduction in production batch size:

$$\sqrt{\frac{2DC}{(1-D/P)UF}} - \sqrt{\frac{2DC}{(1-D/P)U_1F}}$$

$$= \sqrt{\frac{2DC}{(1-D/P)UF}}\left[1-\sqrt{\frac{U}{U_1}}\right]$$

$$= \sqrt{\frac{2DC}{(1-D/P)UF}}\left[1-\sqrt{\frac{1}{(1+M/100)}}\right]$$

And:

$$\% \text{ decrease in } Q = 1-\sqrt{\frac{1}{(1+M/100)}}$$

(ii) Increase in the total cost:

$$U_1D+\sqrt{2DCU_1F(1-D/P)}-UD-\sqrt{2DCUF(1-D/P)}$$

$$=\left\{UD+\sqrt{2DCUF(1-D/P)}\right\}\left[\frac{U_1D+\sqrt{2DCU_1F(1-D/P)}}{UD+\sqrt{2DCUF(1-D/P)}}-1\right]$$

$$=\left\{UD+\sqrt{2DCUF(1-D/P)}\right\}\left[\frac{D(U_1-U)+\sqrt{2DCF(1-D/P)}\left\{\sqrt{U_1}-\sqrt{U}\right\}}{UD+\sqrt{2DCUF(1-D/P)}}\right]$$

$$=\frac{DUM}{100}+\sqrt{2DCUF(1-D/P)}\left\{\sqrt{1+\frac{M}{100}}-1\right\}$$

And:

$$\% \text{ increase in } E = \frac{(DUM/100)+\sqrt{2DCUF(1-D/P)}\left\{\sqrt{(1+M/100)}-1\right\}}{UD+\sqrt{2DCUF(1-D/P)}}$$

Table 5.4 represents the generalised results.

Table 5.4: Result corresponding to %increase in unit production cost.

Reduction in product batch size	$\sqrt{\frac{2DC}{(1-D/P)UF}}\left[1-\sqrt{\frac{1}{(1+M/100)}}\right]$
% decrease in Q	$1-\sqrt{\frac{1}{(1+M/100)}}$
Increase in total cost	$=\frac{DUM}{100}+\sqrt{2DCUF(1-D/P)}\left\{\sqrt{1+\frac{M}{100}}-1\right\}$
% increase in E	$\frac{(DUM/100)+\sqrt{2DCUF(1-D/P)}\left\{\sqrt{(1+M/100)}-1\right\}}{UD+\sqrt{2DCUF(1-D/P)}}$

Refer Tables 5.1 and 5.3. Percentage variation in production batch size is lower in case of higher production cost as compared to that in case of lower production cost. This is because:

$$1-\sqrt{\frac{1}{1+M/100}}<\sqrt{\frac{1}{(1-M/100)}}-1$$

Or $2<\sqrt{\frac{1}{(1-M/100)}}+\sqrt{\frac{1}{(1+M/100)}}$

Or $4<\frac{1}{(1-M/100)}+\frac{1}{(1+M/100)}+\frac{2}{\sqrt{1-(M/100)^2}}$

Or $4<\frac{1+(M/100)+1-(M/100)+2\sqrt{1-(M/100)^2}}{1-(M/100)^2}$

Or $4-4(M/100)^2<2+2\sqrt{1-(M/100)^2}$

Or $2-2(M/100)^2<1+\sqrt{1-(M/100)^2}$

Or $1-2(M/100)^2<\sqrt{1-(M/100)^2}$

Or $1-(M/100)^2<\sqrt{1-(M/100)^2}+(M/100)^2$

Above expression is satisfied also, if:

$$1-(M/100)^2<\sqrt{1-(M/100)^2}$$

Since the practical value of M is between zero and one hundred, the L.H.S. is between 0 and 1. Square root of such a value is greater than that value itself, and therefore the stated expression is true.

With reference to Tables 5.1 and 5.3, percentage variation in total cost appears to be slightly lower in case of higher production cost as compared to that in the case of lower production cost. This is because

$$\frac{(DUM/100)+\sqrt{2DCUF(1-D/P)}\;\left\{\sqrt{(1+M/100)}-1\right\}}{UD+\sqrt{2DCUF(1-D/P)}}<\frac{(DUM/100)+\sqrt{2DCUF(1-D/P)}\;\left\{1-\sqrt{(1-M/100)}\right\}}{UD+\sqrt{2DCUF(1-D/P)}}$$

Or $\sqrt{1+M/100}-1<1-\sqrt{(1-M/100)}$

Or $\sqrt{1+M/100}+\sqrt{(1-M/100)}<2$

Or $1+(M/100)+1-(M/100)+2\sqrt{1-(M/100)^2}<4$

Or $\sqrt{1-(M/100)^2}<1$

As the practical value of M is between zero and one hundred, the above expression is true.

5.3 Stock out inclusion

With inclusion of shortages or stock outs, the output parameters have been provided by Eqs. (1.16), (1.17), and (1.18). Following the similar procedure as explained before, these can be transformed as follows:

$$Q^* = \sqrt{\frac{2DC(K+UF)}{KUF(1-D/P)}} \tag{5.6}$$

$$J^* = \sqrt{\frac{2DCUF(1-D/P)}{K(K+UF)}} \tag{5.7}$$

$$E^* = UD + \sqrt{\frac{2DCKUF(1-D/P)}{(K+UF)}} \tag{5.8}$$

Example 5.4

Consider the following data:

Annual demand, $D = 600$ units

Setup cost, $C =$ ₹ 45

Annual production rate, $P = 960$ units

Unit production cost, $U =$ ₹ 170

Annual shortage cost per unit, $K =$ ₹ 250

Annual inventory carrying cost fraction, $F = 0.2$

Now, from Eq. (5.6):

Optimal production batch size, $Q = 69.36$

From Eq. (5.7):

Optimal maximum shortage quantity, $J = 3.11$

And, from Eq. (5.8):

Optimal total annual cost, $E =$ ₹ 102778.51

In order to accommodate the shortages in production system, batch size has increased with overall total cost reduction. These data can further be used for unit production cost variation, i.e., decrease and increase.

5.3.1 Decreased production cost

With the following information from Example 5:4:

D	C	P	U	K	F	Q	J	E
600	45	960	170	250	0.2	69.36	3.11	102778.51

Unit production cost is reduced as given below:

% decrease in U	5%	10%	15%	20%	25%	30%
U	161.5	153	144.5	136	127.5	119

Table 5.5 shows the effects on different parameters. The observed influence is as follows:

(a) Higher production batch size

(b) Reduced level of shortages

(c) Reduced total cost

Table 5.5: Effects on output parameters corresponding to lower *U* along with shortages.

%Decrease in U	5%	10%	15%	20%	25%	30%
U	161.50	153.00	144.50	136.00	127.50	119.00
Q	70.95	72.68	74.56	76.62	78.89	81.40
% increase in Q	2.29%	4.78%	7.49%	10.46%	13.73%	17.36%
J	3.04	2.97	2.90	2.82	2.74	2.65
% decrease in J	2.24%	4.56%	6.97%	9.47%	12.07%	14.79%
E	97661.08	92543.02	87424.28	82304.81	77184.53	72063.37
% Decrease in E	4.98%	9.96%	14.94%	19.92%	24.90%	29.89%

A generalized approach:

M = % reduction in unit production cost

$$U_1 = \left(1 - \frac{M}{100}\right)U$$

(a) Increase in production batch size:

$$\sqrt{\frac{2DC(K+U_1F)}{KU_1F(1-D/P)}} - \sqrt{\frac{2DC(K+UF)}{KUF(1-D/P)}}$$

$$= \sqrt{\frac{2DC(K+UF)}{KUF(1-D/P)}}\left[\sqrt{\frac{(K+U_1F)U}{(K+UF)U_1}}-1\right]$$

$$= \sqrt{\frac{2DC(K+UF)}{KUF(1-D/P)}}\left[\sqrt{\frac{K+UF(1-M/100)}{(K+UF)(1-M/100)}}-1\right]$$

And:

$$\%\ \text{increase in}\ Q = \sqrt{\frac{K+UF(1-M/100)}{(K+UF)(1-M/100)}}-1$$

(b) Reduction in shortage quantity:

$$\sqrt{\frac{2DCUF(1-D/P)}{K(K+UF)}} - \sqrt{\frac{2DCU_1F(1-D/P)}{K(K+U_1F)}}$$

$$= \sqrt{\frac{2DCUF(1-D/P)}{K(K+UF)}}\left[1-\sqrt{\frac{(K+UF)U_1}{(K+U_1F)U}}\right]$$

$$= \sqrt{\frac{2DCUF(1-D/P)}{K(K+UF)}}\left[1-\sqrt{\frac{(K+UF)(1-M/100)}{K+UF(1-M/100)}}\right]$$

And:

$$\%\ \text{reduction in}\ J = 1-\sqrt{\frac{(K+UF)(1-M/100)}{K+UF(1-M/100)}}$$

(c) Reduction in the total cost:

$$UD+\sqrt{\frac{2DCKUF(1-D/P)}{(K+UF)}} - \left\{U_1D+\sqrt{\frac{2DCKU_1F(1-D/P)}{(K+U_1F)}}\right\}$$

$$= UD+\sqrt{\frac{2DCKUF(1-D/P)}{(K+UF)}} - \left\{1-\frac{U_1D\sqrt{2DCKU_1F(1-D/P)}/K+U_1F}{UD+\sqrt{2DCKUF(1-D/P)}/(K+UF)}\right\}$$

$$= \left\{UD+\sqrt{\frac{2DCKUF\,(1-D/P)}{(K+UF)}}\right\}\left[1-\frac{(1-M/100)UD+\sqrt{2DCK(1-M/100)UF(1-D/P)/\{K+(1-M/100)UF\}}}{UD+\sqrt{2DCKUF(1-D/P)/(K+UF)}}\right]$$

$$= \frac{DUM}{100}+\sqrt{\frac{2CKUF(1-D/P)}{(K+UF)}}\left\{1-\sqrt{\frac{(1-M/100)(K+UF)}{K+(1-M/100)UF}}\right\}$$

And:

$$\% \text{ reduction in } E = \frac{(DUM/100) + \{\sqrt{2DCKUF(1-D/P)/(K+UF)}\}\,[1-\sqrt{(1-M/100)(K+UF)/\{K+(1-M/100)UF\}}]}{UD+\sqrt{2DCKUF(1-D/P)/(K+UF)}}$$

Table 5.6 represents the generalised results.

Table 5.6: Results with shortages corresponding to% reduction in unit production cost.

Increase in production batch size	$\sqrt{\frac{2DC(K+UF)}{KUF(1-D/P)}}\left[\sqrt{\frac{(K+UF)(1-M/100)}{(K+UF)(1-M/100)}}-1\right]$
% increase in Q	$\sqrt{\frac{(K+UF)(1-M/100)}{(K+UF)(1-M/100)}}-1$
Reduction in shortage quantity	$\sqrt{\frac{2DCUF(1-D/P)}{K(K+UF)}}\left[1-\sqrt{\frac{(K+UF)(1-M/100)}{K+UF(1-M/100)}}\right]$
% reduction in J	$1-\sqrt{\frac{(K+UF)(1-M/100)}{K+UF(1-M/100)}}$
Reduction in total cost	$\frac{DUM}{100}+\sqrt{\frac{2DCKUF(1-D/P)}{(K+UF)}}\left\{1-\sqrt{\frac{(1-M/100)(K+UF)}{K+(1-M/100)UF}}\right\}$
% reduction in E	$\frac{(DUM/100)+\{\sqrt{2DCKUF(1-D/P)/(K+UF)}\}\,[1-\sqrt{(1-M/100)(K+UF)/\{K+(1-M/100)UF\}}]}{UD+\sqrt{2DCKUF(1-D/P)/(K+UF)}}$

5.3.2 Increased production cost

With the following information:

D	*C*	*P*	*U*	*K*	*F*	*Q*	*J*	*E*
600	45	960	170	250	0.2	69.36	3.11	102778.51

Unit production cost is reduced as given below:

% increase in U	5%	10%	15%	20%	25%	30%
U	178.50	187.00	195.50	204.00	212.50	221.00

Table 5.7 shows the effects on different parameters. The observed influence is as follows:

(a) Lower production batch size
(b) Relatively increased level of shortages
(c) Increased total cost

Table 5.7: Effects on output parameters corresponding to higher *U* along with shortages.

%Increase in U	5%	10%	15%	20%	25%	30%
U	178.50	187.00	195.50	204.00	212.50	221.00
Q	67.89	66.53	65.26	64.07	62.96	61.92
%decrease in Q	2.12%	4.08%	5.92%	7.63%	9.23%	10.73%
J	3.18	3.25	3.31	3.37	3.43	3.49
%Increase in J	2.16%	4.26%	6.29%	8.26%	10.17%	12.02%
E	107895.36	113011.66	118127.46	123242.78	128357.66	133472.11
% increase in E	4.98%	9.96%	14.93%	19.91%	24.89%	29.86%

A generalised approach:

M = % increase in unit production cost

$$U_1 = U\left(1+\frac{M}{100}\right)$$

(a) Decrease in production batch size:

$$\sqrt{\frac{2DC(K+UF)}{KUF(1-D/P)}} - \sqrt{\frac{2DC(K+U_1F)}{KU_1F(1-D/P)}}$$

$$= \sqrt{\frac{2DC(K+UF)}{KUF(1-D/P)}}\left[1-\sqrt{\frac{(K+U_1F)U}{(K+UF)U_1}}\right]$$

$$= \sqrt{\frac{2DC(K+UF)}{KUF(1-D/P)}}\left[1-\sqrt{\frac{K+UF(1+M/100)}{(K+UF)(1+M/100)}}\right]$$

And:

$$\text{\% decrease in } Q = 1-\sqrt{\frac{K+UF(1+M/100)}{(K+UF)(1+M/100)}}$$

(b) Increase in shortage quantity:

$$\sqrt{\frac{2DCU_1F(1-D/P)}{K(K+U_1F)}}-\sqrt{\frac{2DCUF(1-D/P)}{K(K+UF)}}$$

$$=\sqrt{\frac{2DCUF(1-D/P)}{K(K+UF)}}\left[\sqrt{\frac{(K+UF)U_1}{(K+U_1F)U}}-1\right]$$

$$=\sqrt{\frac{2DCUF(1-D/P)}{K(K+UF)}}\left[\sqrt{\frac{(K+UF)(1+M/100)}{(K+U_1F)(1+M/100)}}-1\right]$$

And:

$$\%\text{ increase in } J=\sqrt{\frac{(K+UF)(1+M/100)}{K+UF(1+M/100)}}-1$$

(c) Increase in the total cost:

$$U_1D+\sqrt{\frac{2DCKU_1F(1-D/P)}{(K+U_1F)}}\left\{UD+\sqrt{\frac{2DCKUF(1-D/P)}{(K+UF)}}\right\}$$

$$=\left\{UD+\sqrt{\frac{2DCKU_1F(1-D/P)}{(K+U_1F)}}\right\}\left[\frac{U_1D+\sqrt{2DCKU_1F(1-D/P)/(K+U_1F)}}{UD+\sqrt{2DCKUF(1-D/P)/(K+UF)}}-1\right]$$

$$=\left\{UD+\sqrt{\frac{2DCKUF(1-D/P)}{(K+UF)}}\right\}\left[\frac{(U_1-U)D+\sqrt{2DCKF(1-D/P)}\left\{\sqrt{U_1/(K+U_1F)}-\sqrt{U/(K+UF)}\right\}}{UD+\sqrt{2DCKUF(1-D/P)/(K+UF)}}\right]$$

$$=\frac{DUM}{100}+\sqrt{\frac{2DCKF(1-D/P)U}{(K+UF)}}\left\{\sqrt{\frac{U_1(K+UF)}{U(K+U_1F)}}-1\right\}$$

$$=\frac{DUM}{100}+\sqrt{\frac{2DCKF(1-D/P)U}{(K+UF)}}\left\{\sqrt{\frac{(1+M/100)(K+UF)}{K+(1+M/100)UF}}-1\right\}$$

And:

$$\%\text{ increase in } E=\frac{(DUM/100)+\{\sqrt{2DCKUF(1-D/P)/(K+UF)}\}[(\sqrt{1+M/100)(K+UF)/\{K+(1-M/100)UF\}}-1]}{UD+\sqrt{2DCKUF(1-D/P)(K+UF)}}$$

Table 5.8 represents the generalised results.

Table 5.8: Results with shortages corresponding to %increase in unit production cost.

Decrease in production batch size	$\sqrt{\frac{2DC(K+UF)}{KUF(1-D/P)}}\left[1-\sqrt{\frac{K+UF(1+M/100)}{(K+UF)(1+M/100)}}\right]$
% decrease in Q	$1-\sqrt{\frac{K+UF(1+M/100)}{(K+UF)(1+M/100)}}$
Increase in shortage quantity	$\sqrt{\frac{2DUF(1-D/P)}{K(K+UF)}}\left[\sqrt{\frac{K+UF(1+M/100)}{(K+UF)(1+M/100)}}-1\right]$
% reduction in J	$\sqrt{\frac{(K+UF)(1+M/100)}{K+UF(1+M/100)}}-1$
Increase in total cost	$\frac{DUM}{100}+\sqrt{\frac{2DCKUF(1-D/P)}{(K+UF)}}\left\{\sqrt{\frac{(1+M/100)(K+UF)}{K+(1+M/100)UF}}-1\right\}$
% increase in E	$\frac{(DUM/100)+\{\sqrt{2DCKUF(1-D/P)/(K+UF)}\}\left[\sqrt{(1+M/100)(K+UF)/\{K+(1+M/100)UF\}}-1\right]}{UD+\sqrt{2DCKUF(1-D/P)/(K+UF)}}$

5.4 Interaction of production cost with other parameters

This interaction can be in the context of lower and higher production cost.

5.4.1 Production cost decrease

As discussed before, production batch size increases with a reduction in production cost. A suitable option can be the set up cost reduction with an aim of similar batch size.

Example 5.5

Consider the following information:

D	*C*	*P*	*U*	*F*	*Q*	*E*
600	45	960	170	0.2	65.08	102829.76

With a reduced production cost of ₹ 153, optimal batch quantity increases to 68.60 units.

For a similar batch quantity, a reduced setup cost ($C1$) can be obtained as follows:

$$65.08 = \sqrt{\frac{2 \times 600 \times C_1}{(1 - 600/900) \times 153 \times 0.2}}$$

Or $$C_1 = ₹\,40.50$$

When a change is triggered by an unavoidable demand reduction, batch quantity reduces.

In order to have similar batch quantity, a potentially reduced production cost can be a response.

With a decreased demand of 570 units, the batch quantity is 60.94. For the similar batch size, a potentially reduced production cost $U1$ is obtained as follows:

$$65.08 = \sqrt{\frac{2 \times 570 \times 45}{(1 - 570/900) \times U_1 \times 0.2}}$$

Or $$U_1 = ₹149.07$$

In order to generalise:

$$D_1 = D\left(1 - \frac{M}{100}\right)$$

$$U_1 = U\left(1 - \frac{N}{100}\right)$$

$$\sqrt{\frac{2DC}{(1 - D/P)UF}} = \sqrt{\frac{2D_1C}{(1 - D_1/P)U_1F}}$$

Or $$\frac{D}{(1 - D/P)U} = \frac{D_1}{(1 - D_1/P)U_1}$$

Or $$\frac{U_1}{U} = \frac{D_1(1 - D/P)}{D(1 - D_1P)}$$

Or $$1 - \frac{N}{100} = \frac{(1 - M/100)(1 - D/P)}{(1 - D_1/P)}$$

Or $$\frac{N}{100} = 1 - \frac{(1 - M/100)(1 - D/P)}{(1 - D_1/P)}$$

Or $$\frac{N}{100} = \frac{1 - (D_1/P) - (1 - M/100)(1 - D/P)}{1 - D(1 - M/100)/P}$$

Or $$\frac{N}{100} = \frac{1 - (D_1/P) - (1 - D/P) + (M/100)(1 - D/P)}{1 - (D/P)(1 - M/100)}$$

$$\text{Or} \quad \frac{N}{100} = \frac{1-(D/P)-(1-D_1/P)+(M/100)(1-D/P)}{1-(D/P)(1-M/100)}$$

$$\text{Or} \quad \frac{N}{100} = \frac{(D/P)-(M/100)+(M/100)(1-D/P)}{1-(D/P)(1-M/100)}$$

$$\text{Or} \quad \frac{N}{100} = \frac{(M/100)}{1-(D/P)(1-M/100)}$$

$$\text{Or} \quad N = \frac{M}{1-(D/P)(1-M/100)}$$

For the present case, values of *N* are provided in Table 5.9 corresponding to a given value of *M*. Values of *N* are higher than that of *M* and less sensitive towards higher values of *M*.

Table 5.9: Variation of *N* (production cost) with respect to *M* (demand).

S. No.	M	$N = \frac{M}{1-(D/P)(1-M/100)}$
1	1	2.62
2	3	7.62
3	5	12.31
4	7	16.72
5	9	20.87

5.4.2 Production cost increase

As discussed before, production batch size decreases with an increase in production cost.

A suitable option can be the reduction in *F* with an aim of similar batch size.

Example 5.6

Consider the following information:

D	C	P	U	F	Q	E
600	45	960	170	0.2	65.08	102829.76

With an increased production cost of ₹ 178.5, optimal batch quantity reduces to 63.51 units.

For a similar batch quantity, a reduced value of *F*, i.e., F_1 can be obtained as follows:

$$65.08 = \sqrt{\frac{2 \times 600 \times 45}{(1 - 600/960) \times 178.5 \times F_1}}$$

Or $$F_1 = 0.19$$

In order to generalise:

$$U_1 = U\left(1 + \frac{M}{100}\right)$$

$$F_1 = F\left(1 - \frac{N}{100}\right)$$

$$\sqrt{\frac{2DC}{(1 - D/P)UF}} = \sqrt{\frac{2DC}{(1 - D/P)U_1F_1}}$$

Or $$U_1F_1 = UF$$

Or $$F_1 = \frac{UF}{U_1}$$

Or $$\frac{1 - N}{100} = \frac{1}{(1 + M/100)}$$

Or $$\frac{N}{100} = 1 - \frac{1}{(1 + M/100)}$$

Or $$\frac{N}{100} = \frac{M/100}{(1 + M/100)}$$

Or $$N = \frac{M}{(1 + M/100)}$$

For the present case, values of *N* are provided in Table 5.10 corresponding to a given value of *M*. Values of *N* are lower than that of *M* and also less sensitive towards higher values of *M*.

Table 5.10: Variation of *N* (*F*) with respect to *M* (*U*).

S. No.	*M*	$N = \frac{M}{(1 + M/100)}$
1	5	4.76
2	10	9.09
3	15	13.04
4	20	16.67
5	25	20

6
Production Rate

Abstract : The number of products concerning a finished item manufactured per unit time is referred to as a production rate. This also gives an idea for availability of a product in terms of numbers per unit time related to the completion of processing at a group of facilities finally. Concept of production rate has been discussed along with the production rate increase and decrease. A good knowledge of potential effects helps the management in case of such situations that may occur in practice. Interaction of production rate with other parameters is analysed in this chapter and also illustrated with examples.

Keyword : Production rate, production rate increase, production rate decrease, potential effects, examples

In case of one facility (as shown in Fig. 6.1), manufacturing rate or production rate needs to be estimated for furthermore analysis.

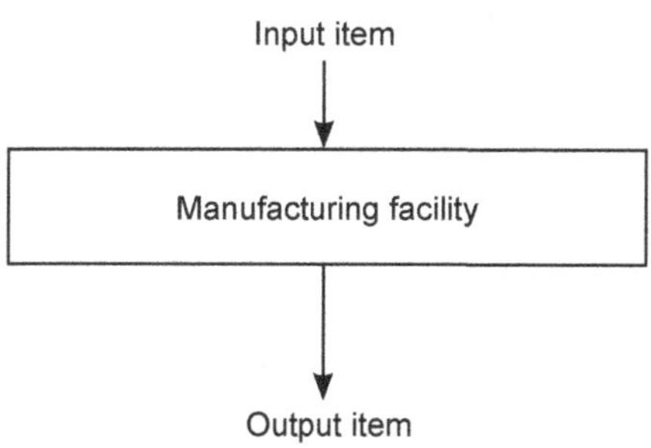

Figure 6.1: Facility production

Input items or raw material is processed at a manufacturing facility. After the processing, output item or finished product is achieved at a certain rate. Number of products concerning finished item manufactured per unit time is referred to as a production rate. In case where a group of different facilities (as shown in Fig. 6.2) is used for manufacturing a finished item finally, an overall production rate can be estimated. This gives an idea for availability of a product in terms of numbers per unit time related to the completion of processing at a group of facilities finally.

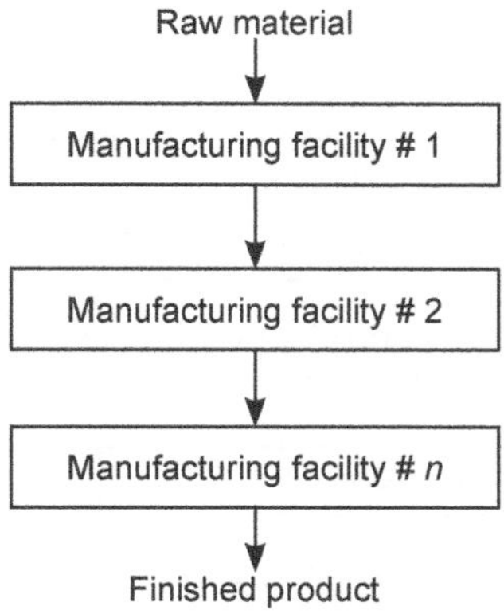

Figure 6.2: Production at a group of facilities

6.1 Production rate increase

Different factors/reasons for an increase in production rate are as follows:

(a) Because of an enhanced degree of automation, production rate might increase.

(b) Material may be substituted by relatively light material, contributing towards faster handing within the plant. This results into an overall higher production rate.

(c) While observing the operational process, it may be noticed there is certain level of manual effort in packaging operation. If number of persons can be increased for relevant operation, an overall production rate for finally packaged product can be enhanced.

(d) While elimination production inefficiencies at various, stages, potential benefit of an increased production rate might be achieved.

In general, the management makes efforts from time to time, in order to increase the rate of production and therefore the analysis may include this. With the increased production rate (say P_1), the production batch size reduces and facility is to be run for shorter period. Fig. 6.3 represents such case.

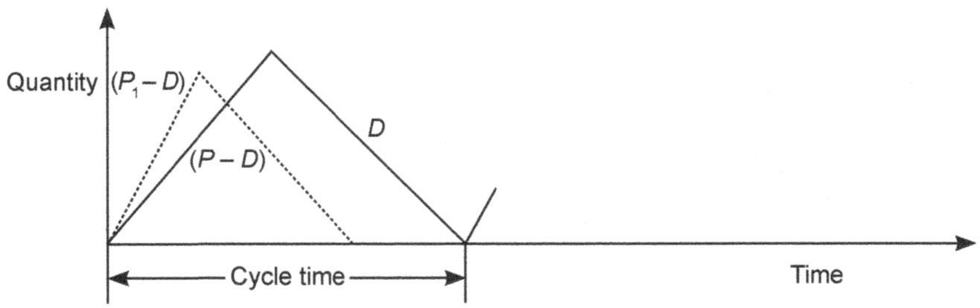

Figure 6.3: Production rate increase

Example 6.1

Consider the parameters as follows.

D	*C*	*I*	*P*	*Q*	*E*
600	45	40	960	60.00	900.00

Implement an increase in production rate as give below:

% Increase in P	5%	10%	15%	20%	25%	30%
P	1008	1056	1104	1152	1200	1248

Table 6.1 shows the concerned effects. It can be observed that:

(i) Optimal production batch size reduces.

(ii) Total related batch size reduces.

Table 6.1: Influence of production rate increase on parameters.

% Increase in P	5%	10%	15%	20%	25%	30%
P	1008	1056	1104	1152	1200	1248
Q	57.75	55.91	54.38	53.02	51.96	50.99
% Decrease in Q	3.75%	6.81%	9.37%	11.53%	13.40%	15.02%
E	9354.03	965.78	993.02	1017.35	1039.23	1059.23
% increase in E	3.89%	7.31%	7.31%	10.34%	15.47%	17.67%

For a general approach:

M = % increase in production rate

$$P_1 = P\left(1+\frac{M}{100}\right)$$

(a) Reduction in production batch size:

$$\sqrt{\frac{2DC}{(1-D/P)I}} - \sqrt{\frac{2DC}{(1-D/P_1)I}}$$

$$= \sqrt{\frac{2DC}{(1-D/P)I}}\left[1-\sqrt{\frac{(1-D/P)}{(1-D/P_1)}}\right]$$

$$= \sqrt{\frac{2DC}{(1-D/P)I}}\left[1-\sqrt{\frac{(1-D/P)}{1\{D/P(1+M/100)\}}}\right]$$

And:

$$\% \text{ reduction in } Q = 1 - \sqrt{\frac{(1 - D/P)}{1\{D/P(1 + M/100)\}}}$$

(b) Increase in total cost:

$$\sqrt{2DCI(1 - D/P_1)} - \sqrt{2DCI(1 - D/P)}$$

$$= \sqrt{2DCI(1 - D/P)}\left[\sqrt{\frac{1 - D/P_1}{(1 - D/P)}} - 1\right]$$

$$= \sqrt{2DCI(1 - D/P)}\left[\sqrt{\frac{1 - \{D/P(1 + M/100)\}}{(1 - D/P)}} - 1\right]$$

And:

$$\% \text{ increase in } E = \sqrt{\frac{1 - \{D/P(1 + M/100)\}}{(1 - D/P)}} - 1$$

Tables 6.2 summarises these results.

Table 6.2: Results with respect to % production rate increase.

Reduction in the production batch size	$\sqrt{\frac{2DC}{(1 - D/P)I}}\left[1 - \sqrt{\frac{(1 - D/P)}{1 - \{D/P(1 + M/100)\}}}\right]$
% reduction in the production batch size	$1 - \sqrt{\frac{(1 - D/P)}{1 - \{D/P(1 + M/100)\}}}$
Increase in the total cost	$\sqrt{2DCI(1 - D/P)}\left[\sqrt{\frac{1 - \{D/P(1 + M/100)\}}{(1 - D/P)}} - 1\right]$
% Increase in the total cost	$\sqrt{\frac{1 - \{D/P(1 + M/100)\}}{(1 - D/P)}} - 1$

6.2 Production rate decrease

Various factors for a reduction in the production rate are given below:

(a) Agricultural workforce is employed for certain portion of unskilled or less skilled jobs/tasks. Because of agricultural season, there is sudden drop in number of human resources. As a result of this, overall production rate goes down.

(b) In case where shelf life is an issue, enough stock might be available and therefore the production rate needs to be reduced.

(c) Because of less experience, human resources might work at a comparatively slow pace and therefore an overall production rate goes down.

(d) A lower production rate may also be the result of frequent maintenance problems or breakdown.

Knowledge of potential effects helps the management in case of such situation, i.e., a production rate decrease. Fig. 6.4 also represents the reduction in rate of production. With the decreased production rate (say $P1$), the production batch size increases and facility is to be run for longer period.

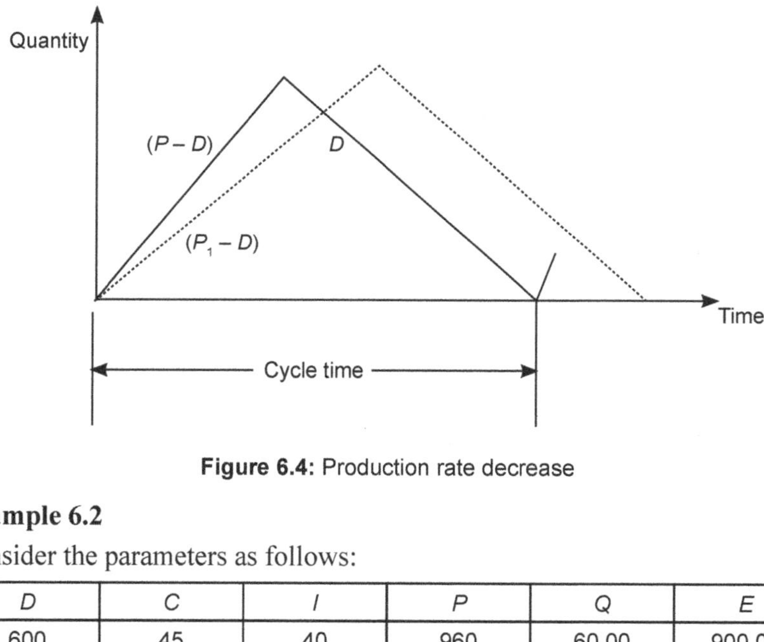

Figure 6.4: Production rate decrease

Example 6.2

Consider the parameters as follows:

D	*C*	*I*	*P*	*Q*	*E*
600	45	40	960	60.00	900.00

Implement a reduction in production rate as given below:

% Decrease in P	5%	10%	15%	20%	25%	30%
P	912	864	816	768	720	672

Table 6.3 shows the concerned effects. It can be observed that:

(i) Optimal production batch size increases.

(ii) Total related cost decreases.

Table 6.3: Influence of production rate decrease on parameters.

% decrease in P	5%	10%	15%	20%	25%	30%
P	912	864	816	768	720	672
Q	62.82	66.47	71.41	78.56	90.00	112.25
% Increase in Q	4.70%	10.78%	19.02%	30.93%	50.00%	87.08%
E	859.62	812.40	756.15	687.39	600.00	481.07
% Decrease in E	4.49%	9.73%	15.98%	23.62%	33.33%	46.55%

For a general approach:

M = % decrease in production rate

$$P_1 = P\left(1-\frac{M}{100}\right)$$

(a) Increase in production rate

$$\sqrt{\frac{2DC}{(1-D/P_1)I}}-\sqrt{\frac{2DC}{(1-D/P)I}}$$

$$=\sqrt{\frac{2DC}{(1-D/P)I}}\left[\sqrt{\frac{(1-D/P)}{(1-D/P_1)}}-1\right]$$

$$=\sqrt{\frac{2DC}{(1-D/P)I}}\left[\sqrt{\frac{(1-D/P)}{1-\{D/P(1-M/100)\}}}-1\right]$$

And:

$$\%\text{ increase in } Q=\sqrt{\frac{(1-D/P)}{1-\{D/P(1-M/100)\}}}-1$$

(b) Reduction in total cost:

$$\sqrt{2DCI(1-D/P)}-\sqrt{2DCI(1-D/P_1)}$$

$$=\sqrt{2DCI(1-D/P)}\left[1-\sqrt{\frac{(1-D/P_1)}{(1-D/P)}}\right]$$

$$=\sqrt{2DCI(1-D/P)}\left[1-\sqrt{\frac{1-\{D/P(1-M/100)\}}{(1-D/P)}}\right]$$

And:

$$\% \text{ reduction in } E = 1 - \sqrt{\frac{1-\{D/P(1-M/100)\}}{(1-D/P)}}$$

Table 6.4 summarizes these results.

Table 6.4: Results with respect to %production rate decrease.

Increase in the production batch size	$\sqrt{\frac{2DC}{(1-D/P)I}}\left[\sqrt{\frac{(1-D/P)}{1-\{D/P(1-M/100)\}}}-1\right]$
% increase in the production batch size	$\sqrt{\frac{(1-D/P)}{1-\{D/P(1-M/100)\}}}-1$
Reduction in the total cost	$\sqrt{2DCI(1-D/P)}\left[1-\sqrt{\frac{1-\{D/P(1-M/100)\}}{(1-D/P)}}\right]$
% reduction in the total cost	$1-\sqrt{\frac{1-\{D/P(1-M/100)\}}{(1-D/P)}}$

6.3 Including shortages

When shortages are incorporated in the manufacturing system, the related expressions are as follows:

$$Q^* = \sqrt{\frac{2DC(K+1)}{KI(1-D/P)}}$$

$$J^* = \sqrt{\frac{2DCI(1-D/P)}{K(K+I)}}$$

$$E^* = \sqrt{\frac{2DCKI(1-D/P)}{(K+I)}}$$

With the inclusion of shortages, variation in rate of production is implemented and analyzed.

6.3.1 Higher production rate

When production rate is increased from P to P_1, it is depicted by Fig. 6.5. It is of interest to know the influence on:

(i) Production batch size
(ii) Shortage quantity
(iii) Total related cost

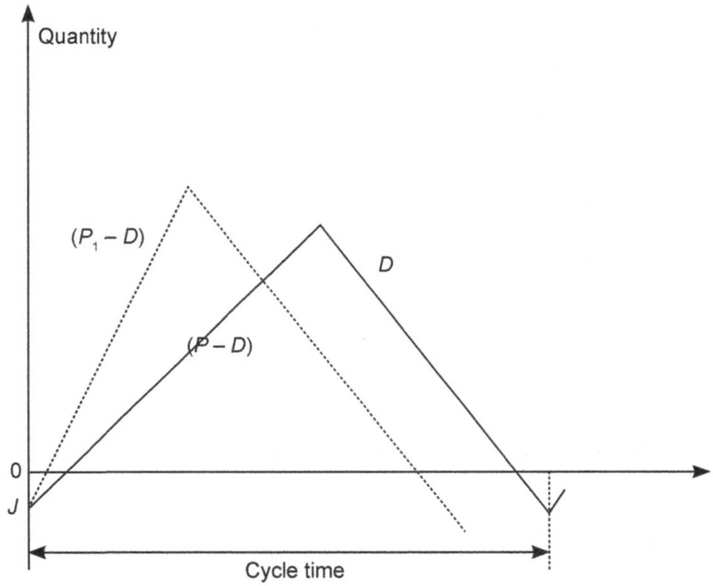

Figure 6.5: Higher production rate with shortages

Effects of the higher production rate are illustrated with the following Example.

Example 6.3

With the available information:

P	*D*	*C*	*I*	*K*	*Q*	*E*	*J*
960	600	45	40	100	71.0	760.64	7.61

Implement an increased production rate as given below:

% increase in P	5%	10%	15%	20%	25%	30%
P	1008	1056	1104	1152	1200	1248

Table 6.5 shows the concerned effects. It can be observed that:

(i) Optimal production batch size reduces.
(ii) Total related cost increases.
(iii) Optimal shortage quantity increases

Table 6.5: Influence of production rate increase on parameters along with shortages.

%Increase in P	5%	10%	15%	20%	25%	30%
P	1008	1056	1104	1152	1200	1248
Q	68.33	66.16	64.34	62.80	61.48	60.33
% Decrease in Q	3.75%	6.81%	9.37%	11.53%	13.40%	15.02%
E	790.25	816.23	839.25	859.82	878.31	895.04
%Increase in E	3.89%	7.31%	10.34%	13.04%	15.47%	17.67%
J	7.90	8.16	8.39	8.60	8.78	8.95
% increase in J	3.89%	7.31%	10.34%	13.04%	15.47%	17.67%

For a general approach:

$$P_1 = P\left(1+\frac{M}{100}\right)$$

(a) Reduction in the production batch size

$$= \sqrt{\frac{2DC(K+I)}{KI(1-D/P)}} - \sqrt{\frac{2DC(K+I)}{KI(1-D/P_1)}}$$

$$= \sqrt{\frac{2DC(K+I)}{KI(1-D/P)}}\left[1-\sqrt{\frac{(1-D/P)}{(1-D/P_1)}}\right]$$

$$= \sqrt{\frac{2DC(K+I)}{KI(1-D/P)}}\left[1-\sqrt{\frac{(1-D/P)}{1-\{D/P(1+M/100)\}}}\right]$$

And:

$$\%\text{reduction in } Q = 1-\sqrt{\frac{(1-D/P)}{1-\{D/P(1+M/100)\}}}$$

(b) Increase in the total cost $= \sqrt{\frac{2DCKI(1-D/P_1)}{(K+I)}} - \sqrt{\frac{2DCKI(1-D/P)}{(K+I)}}$

$$= \sqrt{\frac{2DCKI(1-D/P)}{(K+I)}}\left[\sqrt{\frac{(1-D/P_1)}{(1-D/P)}}-1\right]$$

$$= \sqrt{\frac{2DCKI(1-D/P)}{(K+I)}}\left[\sqrt{\frac{1-\{D/P(1+M/100)\}}{(1-D/P)}}-1\right]$$

And:

$$\% \text{ increase in } E = \sqrt{\frac{1-\{D/P(1+M/100)\}}{(1-D/P)}} - 1$$

(c) Increase in the maximum shortage quantity

$$= \sqrt{\frac{2DCI(1-D/P_1)}{K(K+I)}} - \sqrt{\frac{2DCI(1-D/P)}{K(K+I)}}$$

$$= \sqrt{\frac{2DCI(1-D/P)}{K(K+I)}} \left[\sqrt{\frac{(1-D/P_1)}{(1-D/P)}} - 1 \right]$$

$$= \sqrt{\frac{2DCI(1-D/P)}{K(K+I)}} \left[\sqrt{\frac{1-\{D/P(1+M/100)\}}{(1-D/P)}} - 1 \right]$$

And:

$$\% \text{ increase in } J = \sqrt{\frac{1-\{D/P(1+M/100)\}}{(1-D/P)}} - 1$$

Table 6.6 summarizes the derived results.

Table 6.6: Result with respect to % production rate increase including shortages.

Reduction in the production batch size	$\sqrt{\frac{2DC(K+I)}{KI(1-D/P)}} \left[1 - \frac{(1-D/P)}{1-\{D/P(1+M/100)\}} \right]$
% reduction in the production batch size	$1 - \sqrt{\frac{(1-D/P)}{1-\{D/P(1+M/100)\}}}$
Increase in the total cost	$\sqrt{\frac{2DCKI(1-D/P)}{(K+I)}} \left[\sqrt{\frac{1-\{D/P(1+M/100)\}}{(1-D/P)}} - 1 \right]$
% increase in the total cost	$\sqrt{\frac{1-\{D/P(1+M/100)\}}{(1-D/P)}} - 1$
Increase in the maximum shortage quantity	$\sqrt{\frac{2DCI(1-D/P)}{K(K+I)}} \left[\sqrt{\frac{1-\{D/P(1+M/100)\}}{(1-D/P)}} - 1 \right]$
% increase in J	$\sqrt{\frac{1-\{D/P(1+M/100)\}}{(1-D/P)}} - 1$

6.3.2 Lower production rate

When production rate is lowered from P to P_1, it is depicted by Fig. 6.6. As it will be demonstrated with the help of an example, the following effects are observed:

(i) Optimal production batch size increases
(ii) Total related cost reduces
(iii) Optimal shortage quantity reduces

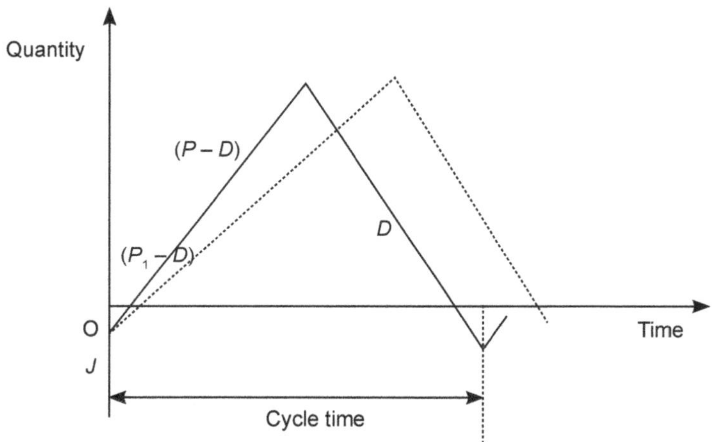

Figure 6.6: Lower production rate with shortages

Example 6.4

With the available information:

P	*D*	*C*	*I*	*K*	*Q*	*E*	*J*
960	600	45	40	100	71.0	760.64	7.61

Implement a lower production rate as given below:

% Decrease in P	5%	10%	15%	20%	25%	30%
P	912	864	816	768	720	672

Table 6.7 shows the concerned effects. Because of an increase in production batch size, enough space needs to be arranged at the final stage as well as between two facilities for work-in-process inventories depending on the specific case. However, total cost reduces and this saving may be utilized for some other activities. In case of limited budget, some other activities including certain developmental projects might get additionally available investment.

Table 6.7: Influence of lower production rate on parameters along with shortages.

% Decrease in P	5%	10%	15%	20%	25%	30%
P	912	864	816	768	720	672
Q	74.33	78.65	84.50	92.95	106.49	132.82
% Increase in Q	4.70%	10.78%	19.02%	30.93%	50.00%	87.08%
E	726.51	686.61	639.06	580.95	507.09	406.58
% decrease in E	4.49%	9.73%	15.98%	23.62%	33.33%	46.55%
J	7.27	6.87	6.39	5.81	5.07	4.07
% Decrease in J	4.49%	9.73%	15.98%	23.62%	33.33%	46.55%

For a general approach:

$$P_1 = \left(1 - \frac{M}{100}\right)P$$

(a) Increase in the production batch size

$$= \sqrt{\frac{2DC(K+I)}{KI(1-D/P_1)}} - \sqrt{\frac{2DC(K+I)}{KI(1-D/P)}}$$

$$= \sqrt{\frac{2DC(K+I)}{KI(1-D/P)}}\left[\sqrt{\frac{(1-D/P)}{(1-D/P_1)}} - 1\right]$$

$$= \sqrt{\frac{2DC(K+I)}{KI(1-D/P)}}\left[\sqrt{\frac{(1-D/P)}{1-\{D/P(1-M/100)\}}} - 1\right]$$

And:

$$\%\text{ increase in } Q = \sqrt{\frac{(1-D/P)}{1-\{D/P(1-M/100)\}}} - 1$$

(b) Reduction in total cost $= \sqrt{\frac{2DCKI(1-D/P)}{(K+I)}} - \sqrt{\frac{2DCKI(1-D/P_1)}{(K+I)}}$

$$= \sqrt{\frac{2DCKI(1-D/P)}{(K+I)}}\left[1 - \sqrt{\frac{(1-D/P_1)}{(1-D/P)}}\right]$$

$$= \sqrt{\frac{2DCKI(1-D/P)}{(K+I)}}\left[1 - \sqrt{\frac{1-\{D/P(1-M/100)\}}{(1-D/P)}}\right]$$

And:

$$\%\text{ reduction in } E = 1 - \sqrt{\frac{1-\{D/P(1-M/100)\}}{(1-D/P)}}$$

(c) Reduction in the maximum shortage quantity

$$= \sqrt{\frac{2DCI(1-D/P)}{K(K+I)}} - \sqrt{\frac{2DCI(1-D/P_1)}{K(K+I)}}$$

$$= \sqrt{\frac{2DCI(1-D/P)}{K(K+I)}}\left[1-\sqrt{\frac{(1-D/P_1)}{(1-D/P)}}\right]$$

$$= \sqrt{\frac{2DCI(1-D/P)}{K(K+I)}}\left[1-\sqrt{\frac{1-\{D/P(1-M/100)\}}{(1-D/P)}}\right]$$

And:

$$\%\text{ reduction in } J = 1-\sqrt{\frac{1-\{D/P(1-M/100)\}}{(1-D/P)}}$$

Table 6.8 summarises the derived results.

Table 6.8: Results with respect to % production rate reduction including shortage

Increase in the production batch size	$\sqrt{\frac{2DC(K+I)}{KI(1-D/P)}}\left[\sqrt{\frac{(1-D/P)}{1-\{D/P(1-M/100)\}}}-1\right]$
% increase in the production batch size	$\sqrt{\frac{(1-D/P)}{1-\{D/P(1-100)\}}}-1$
Reduction in the total cost	$\sqrt{\frac{2DCKI(1-D/P)}{(K+I)}}\left[1-\sqrt{\frac{1-\{D/P(1-M/100)\}}{(1-D/P)}}\right]$
% reduction in the total cost	$1-\sqrt{\frac{1-\{D/P(1-M/100)\}}{(1-D/P)}}$
Reduction in the maximum shortage quantity	$\sqrt{\frac{2DCI(1-D/P)}{K(K+I)}}\left[1-\sqrt{\frac{1-\{D/P(1-M/100)\}}{(1-D/P)}}\right]$
% reduction in J	$1-\sqrt{\frac{1-\{D/P(1-M/100)\}}{(1-D/P)}}$

6.4 Interaction of production rate with other parameters

Analytical results are provided in order to examine this interaction of parameters.

6.4.1 Excluding shortages

When the shortages are not considered, then the analysis is carried out for the cases given below:

(i) Change initiated by production rate

(ii) Change triggered by other parameters

6.4.1.1 *Change initiated by the production rate*

(a) Production rate increases

In this case, a suitable response can be potential holding cost reduction for a similar batch size

Now:

M = % increase in production rate

N = % reduction in holding cost

$$P_1 = \left(1 + \frac{M}{100}\right)P$$

$$I_1 = \left(1 - \frac{N}{100}\right)I$$

For similar batch size:

$$\sqrt{\frac{2DC}{(1-D/P)I}} = \sqrt{\frac{2DC}{(1-D/P_1)I_1}}$$

Or
$$(1 - D/P_1)I_1 = (1 - D/P)I$$

Or
$$\frac{I_1}{I} = \frac{(1-D/P)}{(1-D/P_1)}$$

Or
$$1-\frac{N}{100} = \frac{(1-D/P)}{(1-D/P_1)}$$

Or
$$\frac{N}{100} = 1-\frac{(1-D/P)}{(1-D/P_1)}$$

Or
$$\frac{N}{100} = \frac{1-(D/P_1)-1+(D/P)}{(1-D/P_1)}$$

Or
$$\frac{N}{100} = \frac{(D/P)-\{D/P(1+M/100)\}}{1-\{D/P(1+M/100)\}}$$

Or
$$\frac{N}{100} = \frac{(1+M/100)(D/P)-(D/P)}{(1+M/100)-(D/P)}$$

Or
$$\frac{N}{100} = \frac{(D/P)(M/100)}{(1-D/P)+(M/100)}$$

Or

$$N = \frac{M(D/P)}{(1-D/P)+(M/100)}$$

With the available information:

D	C	I	P	Q	E
600	45	40	960	60.00	900.00

And with the use of relevant data, values of N corresponding to different values of M are provided in Table 6.9. Values of N are less sensitive towards higher values of M.

Table 6.9: Variation of N (holding cost) with respect to M (production rate).

S. No.	M	$N = \frac{M(D/P)}{(1-D/P)+(M/100)}$
1	10	13.16
2	15	17.86
3	20	21.74
4	25	25
5	30	27.78

(b) Production rate decrease:

Setup cost reduction can be a response for similar batch size.

$$P_1 = \left(1-\frac{M}{100}\right)P$$

$$I_1 = \left(1-\frac{N}{100}\right)I$$

Now:

$$\sqrt{\frac{2DC}{(1-D/P)I}} = \sqrt{\frac{2DC_1}{(1-D/P_1)I_1}}$$

Or

$$\frac{C}{(1-D/P)} = \frac{C_1}{(1-D/P_1)}$$

Or

$$\frac{C_1}{C} = \frac{(1-D/P_1)}{(1-D/P)}$$

Or

$$1-\frac{N}{100} = \frac{(1-D/P_1)}{(1-D/P)}$$

Or

$$\frac{N}{100} = \frac{1-(D/P)-(1-D/P_1)}{(1-D/P)}$$

Or $$\frac{N}{100} = \frac{(D/P_1)-(D/P)}{(1-D/P)}$$

Or $$\frac{N}{100} = \frac{\{D/P(1-M/100)\}-(D/P)}{(1-D/P)}$$

Or $$\frac{N}{100} = \frac{(D/P)\{1-(1-M/100)\}}{(1-M/100)(1-D/P)}$$

Or $$\frac{N}{100} = \frac{(D/P)(M/100)}{(1-M/100)(1-D/P)}$$

Or $$N = \frac{M(D/P)}{(1-M/100)(1-D/P)}$$

With the available information:

D	*C*	*I*	*P*	*Q*	*E*
600	45	40	960	60.00	900.00

And with the use of relevant data, values of N corresponding to different values of *M* are provided in Table 6.10. Corresponding values of *N* are more than of *M*. Furthermore, these values of N are more sensitive towards higher values of M, and therefore comparatively greater effort might be required in order to reduce the setup cost for higher decrease in production rate.

Table 6.10: Variation of *N* (setup cost) with respect to *M* (production rate).

S. No.	*M*	$N = \frac{M(D/P)}{(1-M/100)(1-D/P)}$
1	10	18.52
2	15	29.41
3	20	41.67
4	25	55.56
5	30	71.43

6.4.1.2 Change triggered by other parameters

Consider that other parameters including setup/holding cost trigger the variation in output parameters.

(a) Change triggered by setup cost:

(i) Setup cost decrease

Production rate decrease can be a response

$$C_1 = C\left(1-\frac{M}{100}\right)$$

$$P_1 = P\left(1-\frac{N}{100}\right)$$

For similar batch size:

$$\sqrt{\frac{2DC}{(1-D/P)I}} = \sqrt{\frac{2DC_1}{(1-D/P_1)I}}$$

Or $$\frac{C}{(1-D/P)} = \frac{C_1}{(1-D/P_1)}$$

Or $$1-\frac{D}{P_1} = \frac{C_1(1-D/P)}{C}$$

Or $$1-\frac{D}{P_1} = (1-M/100)\,(1-D/P)$$

Or $$\frac{D}{P_1} = 1-(1-M/100)\,(1-D/P)$$

Or $$P_1 = \frac{D}{1-(1-M/100)(1-D/P)}$$

Or $$1-\frac{N}{100} = \frac{(D/P)}{1-(1-M/100)(1-D/P)}$$

Or $$\frac{N}{100} = \frac{1-(1-M/100)(1-D)-(D/P)}{1-(1-M/100)(1-D/P)}$$

Or $$\frac{N}{100} = \frac{1-\{1-(M/100)-(D/P)(1-M/100)\}-(D/P)}{1-\{1-(M/100)-(D/P)(1-M/100)\}}$$

Or $$\frac{N}{100} = \frac{1-1+(M/100)+(D/P)-(D/P)(M/100)-(D/P)}{1-1+(M/100)+(D/P)-(D/P)(M/100)}$$

Or $$\frac{N}{100} = \frac{(M/100)-(D/P)(M/100)}{(M/100)+(D/P)-(D/P)(M/100)}$$

Or $$\frac{N}{100} = \frac{(M/100)(1-D/P)}{(D/P)+(M/100)(1-D/P)}$$

Or $$N = \frac{M(1-D/P)}{(D/P)+(M/100)(1-D/P)}$$

Consider the previous example where:

$D = 600$

$P = 960$

Values of N corresponding to different values of M are provided in Table 6.11. Corresponding values of N are lower than that of M. Furthermore, these values of N are less sensitive towards higher values of M.

Table 6.11: Variation of N (production rate) with respect to M (setup cost reduction).

S. No.	M	$N = \dfrac{M(1-D/P)}{(D/P)+(M/100)(1-D/P)}$
1	10	5.66
2	15	8.26
3	20	10.71
4	25	13.04
5	30	15.25

(ii) Setup cost increase

In case of a specific objective of similar total cost, production rate reduction can be a management response.

$$C_1 = C\left(1+\frac{M}{100}\right)$$

$$P_1 = P\left(1-\frac{N}{100}\right)$$

Now:

$$\sqrt{2DCI(1-D/P)} = \sqrt{2DC_1I(1-D/P_1)}$$

Or $$C(1-D/P) = C_1(1-D/P_1)$$

Or $$1-\frac{D}{P_1} = \frac{C(1-D/P)}{C_1}$$

Or $$1-\frac{D}{P_1} = \frac{(1-D/P)}{(1+M/100)}$$

Or $$\frac{D}{P_1} = \frac{1+(M/100)-1+(D/P)}{(1+M/100)}$$

Or $$\frac{P_1}{D} = \frac{(1+M/100)}{(M/100)+(D/P)}$$

Or $$P\left(1-\frac{N}{100}\right) = \frac{D(1+M/100)}{(M/100)+(D/P)}$$

Or $$1-\frac{N}{100} = \frac{(D/P)(1+M/100)}{(M/100)+(D/P)}$$

Or $$\frac{N}{100} = \frac{(M/100)+(D/P)-(D/P)-(D/P)(M/100)}{(M/100)+(D/P)}$$

Or $$\frac{N}{100} = \frac{(M/100)(1-D/P)}{(M/100)+(D/P)}$$

Or $$N = \frac{M(1-D/P)}{(M/100)+(D/P)}$$

Consider the example where

$D = 600$

$P = 960$

Values of N corresponding to different values of M are provided in Table 6.12.

Corresponding values of N are lower than that of M. Furthermore, these values of N are less sensitive towards higher values of M.

Table 6.12: Variation of N (production rate) with respect to M (setup cost increase).

S. No.	M	$N - \frac{M(1-D/P)}{(M/100)+(D/P)}$
1	10	5.17
2	15	7.26
3	20	9.09
4	25	10.71
5	30	12.16

(b) Change triggered by holding cost:

(i) Holding cost reduction

Production rate increase can be a response for similar batch size objective.

$$I_1 = I\left(1-\frac{M}{100}\right)$$

$$P_1 = P\left(1+\frac{N}{100}\right)$$

Now:

$$\sqrt{\frac{2DC}{(1-D/P)I}} = \sqrt{\frac{2DC}{(1-D/P_1)I_1}}$$

Or $(1 - D/P_1)I_1 = (1 - D/P)\,I$

Or $(1 - D/P_1)(1 - M/100) = (1 - D/P)$

Or $1 - \frac{D}{P_1} = \frac{(1 - D/P)}{(1 - M/100)}$

Or $\frac{D}{P_1} = \frac{1 - (M/100) - 1 + (D/P)}{(1 - M/100)}$

Or $P_1 = \frac{D(1 - M/100)}{(D/P) - (M/100)}$

Or $1 + \frac{N}{100} = \frac{(D/P)(1 - M/100)}{(D/P) - (M/100)}$

Or $\frac{N}{100} = \frac{(D/P)(1 - M/100) - (D/P) + (M/100)}{(D/P) - (M/100)}$

Or $\frac{N}{100} = \frac{(M/100) - (D/P)(M/100)}{(D/P) - (M/100)}$

Or $N = \frac{M(1 - D/P)}{(D/P) - (M/100)}$

With

$D = 600$

$P = 960$

Values of N corresponding to different values of M are provided in Table 6.13. Values of N are more sensitive towards higher values of M.

Table 6.13: Variation of N (production rate) with respect to M (holding cost reduction)

S. No.	M	$N = \frac{M(1 - D/P)}{(D/P) - (M/100)}$
1	10	7.14
2	15	11.84
3	20	17.65
4	25	25
5	30	34.62

(ii) Holding cost increase

Production rate reduction is a response for similar production batch size.

$$I_1 = I\left(1 + \frac{M}{100}\right)$$

$$P_1 = P\left(1 - \frac{N}{100}\right)$$

Now:

$$\sqrt{\frac{2DC}{(1-D/P)I}} = \sqrt{\frac{2DC}{(1-D/P_1)I_1}}$$

Or $(1 - D/P_1)I_1 = (1 - D/P)I$ (6.1)

Or $(1 - D/P_1)(1 + M/100) = (1 - D/P)$

Or $1 - \frac{D}{P_1} = \frac{(1-D/P)}{(1+M/100)}$

Or $\frac{D}{P_1} = \frac{1+(M/100)-1+(D/P)}{(1+M/100)}$

Or $P_1 = \frac{D(1+M/100)}{(D/P)+(M/100)}$

Or $1 - \frac{N}{100} = \frac{(D/P)(1+M/100)}{(D/P)+(M/100)}$

Or $\frac{N}{100} = \frac{(D/P)+(M/100)-(D/P)-(D/P)(M/100)}{(D/P)+(M/100)}$

Or $\frac{N}{100} = \frac{(M/100)(1-D/P)}{(D/P)+(M/100)}$

Or $\frac{N}{100} = \frac{M(1-D/P)}{(D/P)+(M/100)}$

With:

$D = 600$

$P = 960$

Values of N corresponding to difference values of M are provided in Table 6.14. Values of N are less sensitive towards higher values of M.

Table 6.14: Variation of N (production rate) with respect to M (holding cost increase).

S. No.	M	$N - \frac{M(1-D/P)}{(D/P)+(M/100)}$
1	10	5.17
2	15	7.26
3	20	9.09
4	25	10.71
5	30	12.16

When objective is similar total cost, then also similar result can be obtained. This is

Because:

$$\sqrt{2DCI(1-D/P)} = \sqrt{2DCI_1\ (1-D/P_1)}$$

Or $$I_1(1-D/P_1) = I(1-D/P)$$

That is similar to (6.1)

6.4.2 Including shortages

With the inclusion of shortages also, the analytical result are provided.

6.4.2.1 Variation triggered by production rate

When production rate is increased, than the analysis can be conducted for:

(a) Similar batch size

(b) Similar total cost

These are discussed next.

(a) Similar batch size

Holding cost reduction might be a suitable response. Additionally, if demand has potential to increase, it may also be an option.

(i) Response as holding cost reduction

$$P_1 = P\left(1+\frac{M}{100}\right)$$

$$I_1 = I\left(1-\frac{N}{100}\right)$$

$$\sqrt{\frac{2DC(K+I)}{KI(1-D/P)}} = \sqrt{\frac{2DC(K+I_1)}{KI_1(1-D/P_1)}}$$

Or $$\frac{(K+I)}{I(1-D/P)} = \frac{(K+I_1)}{I_1(1-D/P_1)}$$

Or $$I_1\ (1-D/P_1)(K+I) = I(K+I_1)\ (1-D/P) \qquad (6.2)$$

Or $$(1-N/100)\ (1-D/P_1)(K+I) = K(1-D/P) + I_1\ (1-D/P)$$

Or $$(1-N/100)\ (1-D/P_1)(K+I) - I(1-N/100)(1-D/P) = K(1-D/P)$$

Or $$1-\frac{N}{100} = \frac{K(1-D/P)}{(K+I)(1-D/P_1) - I\ (1-D/P)}$$

Or $\dfrac{N}{100} = \dfrac{(K+I)(1-D/P_1) - I\ (1-D/P) - K(1-D/P)}{(K+I)(1-D/P_1) - I\ (1-D/P)}$

Or $\dfrac{N}{100} = \dfrac{(K+I)(1-D/P_1) - (K+I)(1-D/P)}{K(1-D/P_1) + I(1-D/P_1) - I(1-D/P)}$

Or $\dfrac{N}{100} = \dfrac{(K+I)[1-(D/P_1) - 1 + (D/P)]}{K(1-D/P_1) + I\ [\ 1-(D/P_1) - 1 + (D/P)]}$

Or $\dfrac{N}{100} = \dfrac{(K+I)[(D/P) - (D/P_1)]}{K(1-D/P_1) + I\ [\ (D/P) - (D/P_1)]}$

Or $\dfrac{N}{100} = \dfrac{(K+I)[(D/P) - \{D/P(1+M/100)\}]}{K\{1-D/P(1+M/100)\} + I\ [\ (D/P) - \{D/P(1+M/100)\}]}$

Or $\dfrac{N}{100} = \dfrac{(K+I)\{(1+M/100)(D/P) - (D/P)\}}{K\{(1+M/100) - (D/P)\} + I\{(1+M/100)(D/P) - (D/P)\}}$

Or $\dfrac{N}{100} = \dfrac{(K+I)\{(D/P)(M/100)\}}{K\{(1-D/P) + (M/100)\} + I\{(D/P)(M/100)\}}$

Or $N = \dfrac{M(K+I)(D/P)}{K\{(1-D/P) + (M/100)\} + I\ \{\ (D/P)(M/100)\}}$

With the reference set of data as follows:

P	*D*	*C*	*I*	*K*	*Q*	*E*	*J*
960	600	45	40	100	71	760.64	7.61

And considering the relevant data, the values of *N* are represented by Table 6.15. Values of *N* are higher than that of *M* but are less sensitive towards larger values of *M*.

Table 6.15: Variation of *N* (holding cost) concerning *M* (increased *P*) with shortages.

S. No.	*M*	$N = \dfrac{M(K+I)(D/P)}{K\{(1-D/P) + (M/100)\} + I\{(D/P)(M/100)\}}$
1	2	4.37
2	4	8.24
3	6	11.67
4	8	14.74
5	10	17.50

(ii) Response as demand increase

$$P_1 = P\left(1+\frac{M}{100}\right)$$

$$D_1 = D\left(1+\frac{N}{100}\right)$$

$$\sqrt{\frac{2DC(K+I)}{KI(1-D/P)}} = \sqrt{\frac{2D_1\ C(K+I)}{KI(1-D_1/P_1)}}$$

Or $\dfrac{D}{(1-D/P)} = \dfrac{D_1}{(1-D_1/P_1)}$

Or $D_1(1-D/P) = D(1-D_1/P_1)$

Or $D_1[1-(D/P)+(D/P_1)] = D$

Or $D_1 = \dfrac{D}{(1-(D/P)+(D/P_1)}$

Or $1+\dfrac{N}{100} = \dfrac{1}{1-(D/P)+(D/P_1\ )}$

Or $\dfrac{N}{100} = \dfrac{1-[1-(D/P)+(D/P_1)]}{1-(D/P)+(D/P_1)}$

Or $\dfrac{N}{100} = \dfrac{(D/P)-(D/P_1)}{1-(D/P)+(D/P_1)}$

Or $\dfrac{N}{100} = \dfrac{D(P_1/P)-D}{P_1-D(P_1/P)+D}$

Or $\dfrac{N}{100} = \dfrac{D[(P_1/P)-1]}{P_1-D[(P_1/P)-1]}$

Or $\dfrac{N}{100} = \dfrac{D(M/100)}{P(1+M/100)-D(M/100)}$

Or $N = \dfrac{DM}{P+(P-D)(M/100)}$

With the relevant data such as:

$D = 600$

$P = 960$

The values of N are shown in Table 6.16. These are lower than that of M.

Table 6.16: Variation of *N* (demand) corresponding to *M* (increased *P*) with shortages.

S. No.	*M*	$N = \dfrac{DM}{P + (P - D)(M/100)}$
1	5	3.07
2	10	6.02
3	15	8.88
4	20	11.63
5	25	14.29

(b) Similar total cost

When production rate is increased, the response for similar total cost can be in the form of setup/ holding cost reduction.

(i) Response as setup cost reduction

$$P_1 = P\square\left(1+\frac{M}{100}\right)$$

$$C_1 = C\left(1-\frac{N}{100}\right)$$

Now:

$$\sqrt{\frac{(2DCKI(1-D/P)}{(K+I)}} = \sqrt{\frac{2DC_1\ KI(1-D/P_1)}{(K+I)}}$$

Or $\quad C(1-D/P) = C_1\,(1-D/P_1)$

Or $\quad \dfrac{C_1}{C} = \dfrac{(1-D/P)}{1-D/P_1}$

Or $\quad 1-\dfrac{N}{100} = \dfrac{(1-D/P)}{(1-D/P_1)}$

Or $\quad \dfrac{N}{100} = \dfrac{(1-D/P_1)-1+(D/P)}{1-(D/P_1)}$

Or $\quad \dfrac{N}{100} = \dfrac{(D/P)-(D/P_1)}{1-(D/P_1)}$

Or $\quad \dfrac{N}{100} = \dfrac{D(P_1/P)-D}{P_1-D}$

Or $\quad \dfrac{N}{100} = \dfrac{D(1+M/100)-D}{P(1+M/100)-D}$

Or $$\frac{N}{100} = \frac{(DM/100)}{(P-D)+(PM/100)}$$

Or $$N = \frac{DM}{(P-D)+(PM/100)}$$

With reference to the base data set (for $D = 600$ and $P = 960$), the values of N are shown in Table 6.17. These are observed to be less sensitive towards higher values of M.

Table 6.17: Variation of *N* (setup cost) corresponding to *M* (increased *P*) with shortages.

S. No.	*M*	$N = \frac{DM}{(P-D)+(PM/100)}$
1	5	7.35
2	10	13.16
3	15	17.86
4	20	21.74
5	25	25

(ii) Response as holding cost reduction

$$P_1 = P\left(1+\frac{M}{100}\right)$$

$$I_1 = I\left(1-\frac{N}{100}\right)$$

$$\sqrt{\frac{2DCKI(1-D/P)}{(K+I)}} = \sqrt{\frac{2DCKI_1(1-D/P_1)}{(K+I_1)}}$$

Or $$\frac{I(1-D/P)}{(K+I)} = \frac{I_1(1-D/P_1)}{(K+I_1)}$$

Or $$I_1(1-D/P_1)(K+I) = I(K+I_1)(1-D/P)$$

Since above expression is similar to (6.2), value of N can be obtained as before, i.e.:

$$N = \frac{M(K+I)(D/P)}{K\{1-D/P)+(M/100)\}+I\{(D/P)(M/100)\}}$$

6.4.2.2 Variation triggered by other parameters

Other parameters including demand and holding cost can also initiate the change.

(a) Variation triggered by demand:

(i) Demand increase

Production rate increase can be a response for similar batch size.

$$D_1 = D\left(1+\frac{M}{100}\right)$$

$$P_1 = P\left(1+\frac{N}{100}\right)$$

$$\sqrt{\frac{2DC(K+I)}{KI(1-D/P)}} = \sqrt{\frac{2D_1C(K+I)}{KI(1-D_1/P_1)}}$$

Or $\dfrac{D}{(1-D/P)} = \dfrac{D_1}{(1-D_1/P_1)}$

Or $1-\dfrac{D_1}{P_1} = \dfrac{D_1(1-D/P)}{D}$

Or $1-\dfrac{D_1}{P_1} = (1+M/100)(1-D/P)$

Or $\dfrac{D_1}{P_1} = (1+M/100)(1-D/P)$

Or $P_1 = \dfrac{D_1}{1-(1+M/100)(1-D/P)}$

Or $1+\dfrac{N}{100} = \dfrac{(D_1/P)}{1-(1+M/100)(1-D/P)}$

Or $\dfrac{N}{100} = \dfrac{(D_1/P)-1+(1+M/100)(1-D/P)}{1-(1+M/100)(1-D/P)}$

Or $\dfrac{N}{100} = \dfrac{(D/P)(1+M/100)-1+(1+M/100)-(1+M/100)(D/P)}{1-(1+M/100)+(1+M/100)(D/P)}$

Or $\dfrac{N}{100} = \dfrac{(M/100)}{(1+M/100)(D/P)-(M/100)}$

Or $N = \dfrac{M}{(1+M/100)(D/P)-(M/100)}$

With the reference set of data as follows:

P	*D*	*C*	*I*	*K*	*Q*	*E*	*J*
960	600	45	40	100	71	760.64	7.61

And considering the relevant information, the values of *N* are shown in Table 6.18. Values of *N* are much higher than that of *M*.

Table 6.18: Variation of *N* (production cost) related to *M* (increased *D*) with shortages.

S. No.	M	$N = \dfrac{M}{(1+M/100)(D/P)-(M/100)}$
1	10	17.02
2	15	25.37
3	20	36.36
4	25	47.06
5	30	58.54

(ii) Demand reduction

Production rate reduction is considered as a response.

$$D_1 = D\left(1-\frac{M}{100}\right)$$

$$P_1 = P\left(1-\frac{N}{100}\right)$$

$$\sqrt{\frac{2DC(K+I)}{KI(1-D/P)}} = \sqrt{\frac{2D_1\ C(K+I)}{KI(1-D_1/P_1)}}$$

Or $$\frac{D}{(1-D/P)} = \frac{D_1}{(1-D_1/P_1)}$$

Or $$1-\frac{D_1}{P_1} = \frac{D_1\ (1-D/P)}{D}$$

Or $$1-\frac{D_1}{P_1} = (1-M/100)\ (1-D/P)$$

Or $$\frac{D_1}{P_1} = 1-(1-M/100\ )(1-D/P)$$

Or $$P_1 = \frac{D_1}{1-(1-\ M/100)(1-D/P)}$$

Or $$1-\frac{N}{100} = \frac{(D_1/P)}{1-(1-M/100)(1-D/P)}$$

Or $$\frac{N}{100} = \frac{1-(1-M/100)(1-D/P)(D_1/P)}{1-(1-M/100)(1-D/P)}$$

Or $$\frac{N}{100} = \frac{1-(1-M/100)+(1-M/100)(D/P)-(D/P)(1-M/100\ )}{1-(1-M/100)+(1-M/100)(D/P)}$$

Or $$\frac{N}{100} = \frac{(M/100)}{(M/100\ ) + (1 - M/100)(D/P)}$$

Or $$N = \frac{M}{(M/100) + (1 - M/100)(D/P)}$$

With the reference set of data (including $D = 600$ and $P = 960$), the values of N are shown in Table 6.19. Values of N are higher than that of M.

Table 6.19: Variation of N (production rate) related to M (reduced D) with shortages.

S. No.	M	$N = \frac{M}{(M/100) + (1 - M/100)(D/P)}$
1	10	15.09
2	15	22.02
3	20	28.57
4	25	34.78
5	30	40.68

While comparing with Table 6.18, Values of N are lower. This is because:

$$\frac{M}{(M/100) + (1 - M/100)(D/P)} < \frac{M}{(1 + M/100)(D/P) - (M/100\ )}$$

Or $(M/100\ ) + (1 - M/100)(D/P) > (1 + M/100)(D/P) - (M/100)$

Or $(M/100\ ) + (D/P) - (M/100)\ (D/P) > (D/P) + (M/100)\ (D/P) - (M/100\ )$

Or $(M/50) - (M/50)(D/P) > 0$

Or $(M/50)(1 - D/P) > 0$

Or $(1 - D/P) > 0$

And that is true, because production rate is higher than the demand rate in a batch production environment, in order to analyze production-inventory system.

(b) Variation triggered by holding cost:

(i) Holding cost increase

Production rate decrease can be a response.

$$I_1 = I\left(1 + \frac{M}{100}\right)$$

$$P_1 = P\square\left(1 - \frac{N}{100}\right)$$

For similar batch size:

$$\sqrt{\frac{2DC(K+I)}{KI(1-D/P)}} = \sqrt{\frac{2DC(K+I_1)}{KI_1(1-D/P_1)}}$$

Or $$\frac{(K+I)}{I(1-D/P)} = \frac{(K+I_1)}{I_1(1-D/P_1)} \quad (6.3)$$

Or $$1-\frac{D}{P_1} = \frac{I(K+I_1)(1-D/P)}{I_1(K+I)}$$

Or $$1-\frac{D}{P_1} = \frac{(K+I_1)(1-D/P)}{(1+M/100)(K+I)}$$

Or $$\frac{D}{P_1} = \frac{(1+M/100)(K+I)-(K+I_1)(1-D/P)}{(1+M/100)(K+I)}$$

Or $$P_1 = \frac{D(1+M/100)(K+I)}{(1+M/100)(K+I)-(K+I_1)(1-D/P)}$$

Or $$1-\frac{N}{100} = \frac{(D/P)(1+M/100)(K+I)}{(1+M/100)(K+I)-(K+I_1)(1-D/P)}$$

Or $$\frac{N}{100} = \frac{(1+M/100)(K+I)-(K+I_1)(1-D/P)-(D/P)(1+M/100)(K+I)}{(1+M/100)(K+I)-(K+I_1)(1-D/P)}$$

Or $$\frac{N}{100} = \frac{(1+M/100)(K+I)(1-D/P)-(K+I_1)(1-D/P)}{(1+M/100)(K+I)-[K+I(1+M/100)](1-D/P)}$$

Or $$\frac{N}{100} = \frac{(1-D/P)[K(1+M/100)+I(1+M/100)-K-I(1+M/100)]}{(1+M/100)K+(1+M/100)I-K(1-D/P)-I(1+M/100)(1-D/P)}$$

Or $$\frac{N}{100} = \frac{(1-D/P)(KM/100)}{K[(M/100)+(D/P)]+I(1+M/100)[1-1+(D/P)]}$$

Or $$N = \frac{(1-D/P)KM}{K[(M/100)+(D/P)]+I(1+M/100)(D/P)}$$

Or $$N = \frac{(1-D/P)KM}{(KM/100)+(KD/P)+(ID/P)+(ID/P)(M/100)}$$

Or $$N = \frac{(1-D/P)KM}{(K+I)(D/P)+(M/100)\{K+I(D/P)\}}$$

With the reference set of data as follows:

P	D	C	I	K	Q	E	J
960	600	45	40	100	71	760.64	7.61

And considering the relevant information, the values of N are shown in Table 6.20. These are much lower than that of M and also are less sensitive towards higher values of M.

Table 6.20: Variation of N (production rate) related to M (increased I) with shortages.

S. No.	M	$N = \frac{(1-D/P)KM}{(K+I)(D/P)+(M/100)\{K+I(D/P)\}}$
1	10	3.75
2	15	5.29
3	20	6.67
4	25	7.89
5	30	9

For similar total cost:

$$\sqrt{\frac{2DCKI(1-D/P)}{(K+I)}} = \sqrt{\frac{2DCKI_1(1-D/P_1)}{(K+I_1)}}$$

Or
$$\frac{I(1-D/P)}{(K+I)} = \frac{I_1(1-D/P_1)}{(K+I_1)}$$

Or
$$\frac{(K+I)}{I(1-D/P)} = \frac{(K+I_1)}{I_1(1-D/P_1)}$$

Above expression is similar to (6.3); therefore, similar result can be obtained in this context also.

(ii) Holding cost decrease

Production rate increase can be a response.

$$I_1 = I\left(1-\frac{M}{100}\right)$$

$$P_1 = P\left(1+\frac{N}{100}\right)$$

For similar batch size:

$$\sqrt{\frac{2DC(K+I)}{KI(1-D/P)}} = \sqrt{\frac{2DC(K+I_1)}{KI_1(1-D/P_1)}}$$

Or $$\frac{(K+I)}{I(1-D/P)} = \frac{(K+I_1)}{I_1(1-D/P_1)}$$

Or $$1-\frac{D}{P_1} = \frac{I(K+I_1)(1-D/P)}{I_1\,(K+I)}$$

Or $$1-\frac{D}{P_1} = \frac{(K+I_1)(1-D/P)}{(1-M/100)(K+I)}$$

Or $$\frac{D}{P_1} = \frac{(1-M/100)(K+I)-(K+I_1)(1-D/P)}{(1-M/100)(K+I)}$$

Or $$P_1 = \frac{D(1-M/100)(K+I)}{(1-M/100)(K+I)-(K+I_1)(1-D/P)}$$

Or $$1+\frac{N}{100} = \frac{(D/P)(1-M/100)(K+I)}{(1-M/100)(K+I)-(K+I_1)(1-D/P)}$$

Or $$\frac{N}{100} = \frac{(D/P)(1-M/100)(K+I)-(1-M/100)(K+I)+(K+I_1)(1-D/P)}{(1-M/100)(K+I)-(K+I_1)(1-D/P)}$$

Or $$\frac{N}{100} = \frac{(1-M/100)(K+I)\{(D/P)-1\}+(K+I_1)(1-D/P)}{(1-M/100)(K+I)-[K+I(1-M/100)](1-D/P)}$$

Or $$\frac{N}{100} = \frac{(1-D/P)[K+I(1-M/100)-K(1-M/100)-I(1-M/100)]}{(1-M/100)K+(1-M/100)I-K(1-D/P)-I(1-M/100)(1-D/P)}$$

Or $$\frac{N}{100} = \frac{(1-D/P)(KM/100)}{K[(D/P)-(M/100)]+I(1-M/100)[1-1+(D/P)]}$$

Or $$N = \frac{(1-D/P)\,KM}{[(KD/P)-(KM/100)]+(ID/P)]-(ID/P)(M/100)}$$

Or $$N = \frac{(1-D/P)\,KM}{(K+I)(D/P)-(M/100)\{K+I(D/P)\}}$$

With the reference set of data as follows:

P	D	C	I	K	Q	E	J
960	600	45	40	100	71	760.64	7.61

And considering the relevant information, the values of N are shown in Table 6.21. These are lower than that of M but are more sensitive towards higher values of M.

Table 6.21: Variation of N (production rate) related to M (decreased I) with shortages.

S. No.	M	$N = \frac{(1-D/P)KM}{(K+I)(D/P)-(M/100)\{K+I(D/P)\}}$
1	10	5.00
2	15	8.18
3	20	12.00
4	25	16.67
5	30	22.50

7
Cycle Time

Abstract : A similar production-inventory cycle repeats itself, where cycle time may relate to the period in which certain number of items or quantity is manufactured and consumed completely. There are situations when a cycle time requires to be reduced or increased. After discussing such situations briefly, a focus of the present chapter is on deriving the optimal cycle time and the corresponding total annual cost analytically. Effects of the parameters such as setup cost, production rate, holding cost and demand, on the optimal cycle time are presented in detail. Backordering situation has also been investigated with an inclusion of shortage cost.

Keyword : Production-inventory cycle, cycle time, reduction/increase in cycle time, effects of the input parameters, backordering situation

In the present context, cycle time refers to the duration in which certain number of items or quantity is produced and consumed completely. After this, a similar cycle repeats itself.

7.1 Importance of cycle time

As shown in Fig. 7.1, stock build-up rate is $(P - D)$ where P is the production rate and D is the demand rate. Stock increases during the production time. At the end of the production time, stock starts decreasing. As soon as the stock level becomes zero, the next similar production cycle begins.

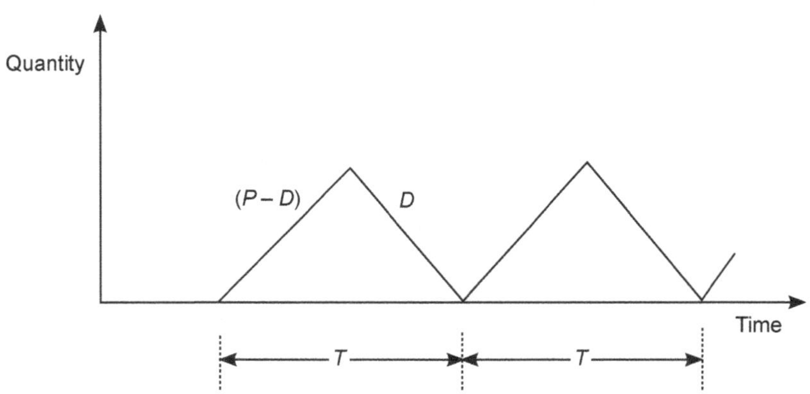

Figure 7.1: Cycle time

Cycle time (T) requires to be reduced with reference to certain optimization when there is:

(a) Reduced facility setup cost
(b) Increased production rate
(c) Increased inventory holding cost
(d) Reduced demand

Furthermore, when shelf life is an issue and product should be consumed timely, the production cycle time might be reduced. These factors are also shown in Fig. 7.2.

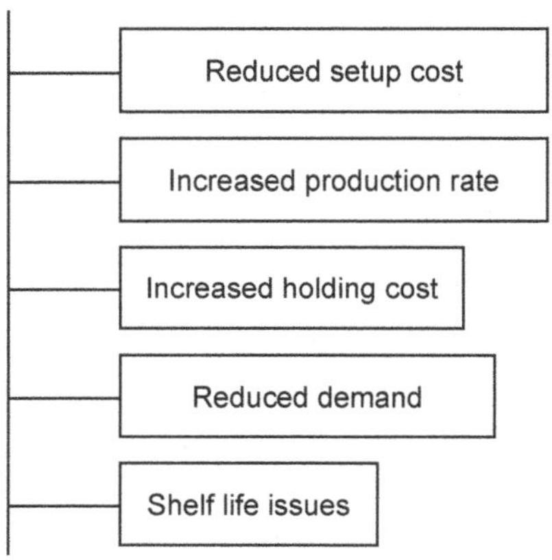

Figure 7.2: Factors for a reduced cycle time

Cycle time (C.T.) may be increased (Fig.7.3) when there is:

(a) Increased facility setup cost
(b) Reduced production rate
(c) Reduced inventory holding cost
(d) Increased demand

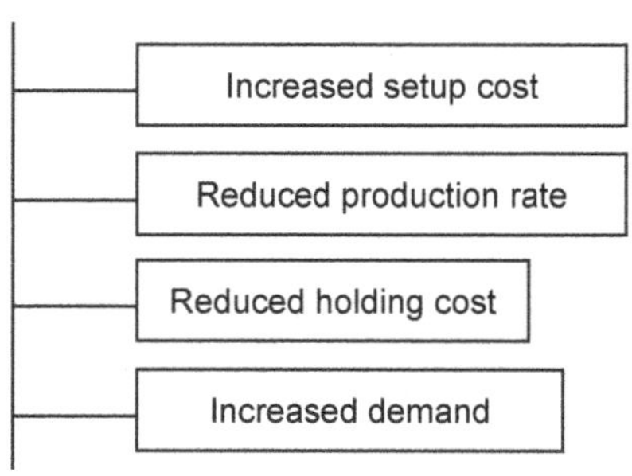

Figure 7.3: Factors for an increased cycle time

Variation in the cycle time (T), i.e., a reduction or increase needs to be studied. For this purpose, the problem should be formulated in terms of T. Now, one cycle of production is shown in Fig. 7.4.

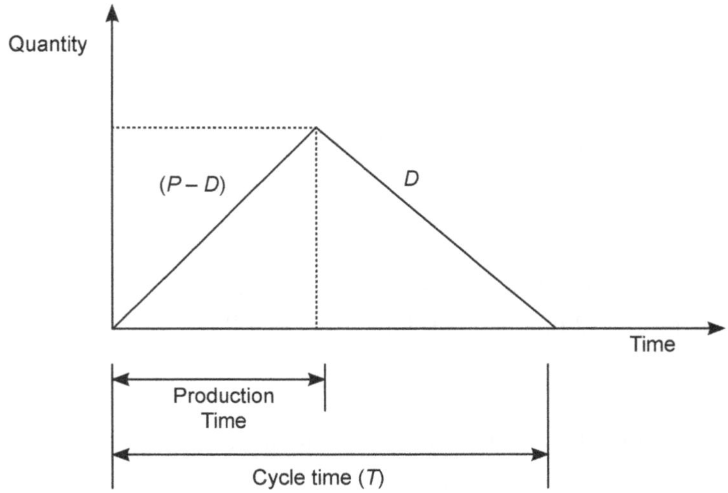

Figure 7.4: A cycle of production

For example, if annual demand (D) is 12,000 products and the cycle time is 0.2 year, then the production quantity per setup is

$$12000 \times 0.2 = 2{,}400$$

Or $12{,}000 \times \frac{1}{5} = 2{,}400$, because there are 5 production cycles in one year.

In order to generalise, production batch size = DT

Production time (P.T.) = $\frac{DT}{P}$

P.T. can also be expressed as:

$$\frac{V}{(P-D)}$$

where V = Maximum stock during the cycle

Now:

$$\frac{V}{(P-D)} = \frac{DT}{P}$$

Or

$$V = \frac{DT(P-D)}{P}$$

Or $$V = DT(1 - D/P)$$

Because average stock is $\left(\frac{V}{2}\right)$, the annual inventory holding cost:

$$AIC = \frac{V}{2}.I$$

Or $$AIC = \frac{DTI\ (1 - D/P)}{2} \tag{7.1}$$

If cycle time (T) is 0.2 year, then the number of facility setup in one year is:

$$\frac{1}{0.2} = 5$$

In order to generalise, number of setup in one year is: $\frac{1}{T}$

And the annual production setup cost is as follows:

$$APC = \frac{1}{T}.C = \frac{C}{T} \tag{7.2}$$

While adding Eqs. (7.1) and (7.2), the total annual cost:

$$E = AIC + APC$$

Or $$E = \frac{DTI(1 - D/P)}{2} + \frac{C}{T} \tag{7.3}$$

In order to obtain the optimal value of T, differentiate w.r.t. T and equate to zero:

$$\frac{DI(1 - D/P)}{2} - \frac{C}{T^2} = 0$$

Or $$\frac{DI(1 - D/P)}{2} = \frac{C}{T^2}$$

Or $$T^2 = \frac{2C}{DI(1 - D/P)}$$

Or $$T^* = \sqrt{\frac{2C}{DI(1 - D/P)}} \tag{7.4}$$

For the optimal of E, substitute the value of $DI(1 - D/P)$ from (7.4) in the Eq. (7.3).

$$E^* = \frac{2C}{T^2}.\frac{T}{2} + \frac{C}{T}$$

Or $$E^* = \frac{C}{T} + \frac{C}{T}$$

Or $$E^* = \frac{2C}{T} \tag{7.5}$$

Example 7.1

Consider the following parameters:

Annual demand, $D = 600$

Annual inventory holding cost per unit, $I = ₹\ 35$

Annual production rate, $P = 960$

Setup cost, $C = ₹\ 50$

From the Eq. (7.4), cycle time:

$$T = \sqrt{\frac{2 \times 50}{600 \times 35(1 - 600/960)}}$$

Or $$T = 0.1127 \text{ year} \approx 0.113 \text{ year}$$

From the Eq. (7.5), the total cost:

$$E = \frac{2 \times 50}{0.1127}$$

Or $$E = ₹\ 887.41$$

7.2 Variation in cycle time

Variation in cycle time can be on lower or higher side.

7.2.1 Lower cycle time

Optimal cycle time decreases because of:

(i) Reduced facility setup cost
(ii) Increased production rate
(iii) Increased holding cost
(iv) Decreased demand

Now:

M = % variation in parameter under consideration

T_1 = Reduced cycle time

(i) Reduced setup cost:

Since the parameter under consideration is setup cost, M refers to the % reduction in setup cost.

Now:

$$C_1 = \left(1 - \frac{M}{100}\right)C$$

$$\text{Decrease in the cycle time} = T - T_1$$

$$= \sqrt{\frac{2C}{DI(1-D/P)}} - \sqrt{\frac{2C_1}{DI(1-D/P)}}$$

$$= \sqrt{\frac{2C}{DI(1-D/P)}}\left[1 - \sqrt{\frac{C_1}{C}}\right]$$

$$= \sqrt{\frac{2C}{DI(1-D/P)}}\left[1 - \sqrt{\left(1 - \frac{M}{100}\right)}\right]$$

And:

$$\%\ \text{decrease in C.T.} = 1 - \sqrt{\left(1 - \frac{M}{100}\right)}$$

In order to illustrate, consider the parameters of Example 7.1 as follows:

D	*C*	*I*	*P*	*T*	*E*
600	50	35	960	0.113	887.41

After implementation of the reduced setup cost, the approximate effects are given as follows:

% decrease in C	5%	10%	15%	20%	25%
C	47.50	45.00	42.50	40.00	37.50
T	0.110	0.107	0.104	0.101	0.098
% decrease in T	2.53%	5.13%	7.80%	10.56%	13.40%
E	864.94	841.87	818.15	793.73	768.52
% decrease in E	2.53%	5.13%	7.80%	10.56%	13.40%

% reduction in the cycle time is lower than that in the facility setup cost.

(ii) Increased production rate

$$P_1 = \left(1 + \frac{M}{100}\right)P$$

$$\text{Decrease in cycle time} = T - T_1$$

$$= \sqrt{\frac{2C}{DI(1-D/P)}} - \sqrt{\frac{2C}{DI(1-D/P_1)}}$$

$$= \sqrt{\frac{2C}{DI(1-D/P)}}\left[1 - \sqrt{\frac{(1-D/P)}{(1-D/P_1)}}\right]$$

$$= \sqrt{\frac{2C}{DI(1-D/P)}}\left[1-\sqrt{\frac{(1-D/P)}{1-\{(D/P(1+M/100)\}}}\right]$$

And:

$$\%\text{ decrease in C.T.} = 1-\sqrt{\frac{(1-D/P)}{1-\{(D/P(1+M/100)\}}}$$

In order to illustrate, consider the reference parameters as follows:

D	*C*	*I*	*P*	*T*	*E*
600	50	35	960	0.113	887.41

After implementation of the production rate, the effects in terms of approximate values are given as follows:

% increase in P	5%	10%	15%	20%	25%	30%
P	1008	1056	1104	1152	1200	1248
T	0.108	0.105	0.102	0.100	0.098	0.096
% decrease in T	3.75%	6.81%	9.37%	11.53%	13.40%	15.02%
E	921.95	952.27	979.13	1003.12	1024.70	1044.21
% increase in E	3.89%	7.31%	10.34%	13.04%	15.47%	17.67%

% decrease in the cycle time is lower in comparison with the % increase in the production rate.

(iii) Increased holding cost

$$I_1 = \left(1+\frac{M}{100}\right)I$$

$$\text{Decrease in cycle time} = T - T_1$$

$$= \sqrt{\frac{2C}{DI(1-D/P)}} - \sqrt{\frac{2C}{DI_1(1-D/P)}}$$

$$= \sqrt{\frac{2C}{DI(1-D/P)}}\left[1-\sqrt{\frac{I}{I_1}}\right]$$

$$= \sqrt{\frac{2C}{DI(1-D/P)}}\left[1-\sqrt{\frac{1}{(1+M/100)}}\right]$$

And:

$$\%\text{ decrease in C.T.} = 1-\sqrt{\frac{1}{(1+M/100)}}$$

In order to illustrate, consider the reference parameters as follows:

D	*C*	*I*	*P*	*T*	*E*
600	50	35	960	0.113	887.41

After implementation of the increased holding cost, the effects in terms of approximate values are provided below:

% increase in *I*	5%	10%	15%	20%	25%
I	36.75	38.50	40.25	42.00	43.75
T	0.110	0.107	0.105	0.103	0.101
% increase in *T*	2.41%	4.65%	6.75%	8.71%	10.56%
E	909.33	930.73	951.64	972.11	992.16
% increase in *E*	2.47%	4.88%	7.24%	9.54%	11.80%

(iv) Decreased demand

$$D_1 = \left(1 - \frac{M}{100}\right) D$$

Decrease in cycle time = $T - T_1$

$$= \sqrt{\frac{2C}{DI(1-D/P)}} - \sqrt{\frac{2C}{D_1 I(1-D_1/P)}}$$

$$= \sqrt{\frac{2C}{DI(1-D/P)}}\left[1 - \sqrt{\frac{D(1-D/P)}{D_1(1-D_1/P)}}\right]$$

$$= \sqrt{\frac{2C}{DI(1-D/P)}}\left[1 - \sqrt{\frac{(1-D/P)}{(1-M/100)\{1-D(1-M/100)/P\}}}\right]$$

And:

$$\text{\% decrease in C.T.} = 1 - \sqrt{\frac{(1-D/P)}{(1-M/100)\{1-D(1-M/100)/P\}}}$$

The reference parameters are as follows:

D	*C*	*I*	*P*	*T*	*E*
600	50	35	960	0.113	887.41

When demand is decreased in this example, the effects in terms of approximate values are given below:

% decrease in D	5%	10%	15%	20%
D	570.00	540.00	510.00	480.00
T	0.111	0.110	0.109	0.109
% decrease in T	1.43%	2.41%	2.99%	3.18%
E	900.26	909.33	914.72	916.52
% increase in E	1.45%	2.47%	3.08%	3.28%

Table 7.1 summarises the generalised results.

Table 7.1: Results for reduced cycle time.

Variation in input parameter	*Decrease in optimal C.T.*
Reduced facility setup cost	Decrease in the cycle time = $\sqrt{\frac{2C}{DI(1-D/P)}}\left[1-\sqrt{\left(1-\frac{M}{100}\right)}\right]$ % decrease in C.T. = $1-\sqrt{\left(1-\frac{M}{100}\right)}$
Increased production rate	Decrease in the cycle time = $\sqrt{\frac{2C}{DI(1-D/P)}}\left[1-\sqrt{\frac{(1-D/P)}{1-\{D/P(1+M/100)\}}}\right]$ % decrease in C.T. = $1-\sqrt{\frac{(1-D/P)}{1-\{D/P(1+M/100)\}}}$
Increased holding cost	Decrease in the cycle time = $\sqrt{\frac{2C}{DI(1-D/P)}}\left[1-\sqrt{\frac{1}{(1+M/100)}}\right]$ % decrease in C.T. = $1-\sqrt{\frac{1}{(1+M/100)}}$
Decreased demand	Decrease in the cycle time = $\sqrt{\frac{2C}{DI(1-D/P)}}\left[1-\sqrt{\frac{(1-D/P)}{(1-M/100)\{1-D(1-M/100)/P\}}}\right]$ % decrease in C.T. = $1-\sqrt{\frac{(1-D/P)}{(1-M/100)\{1-D(1-M/100)/P\}}}$

7.2.2 Higher cycle time

Optimal cycle time increases because of:

(i) Increased facility setup cost
(ii) Decreased production rate
(iii) Decreased holding cost
(iv) Increased demand

Now:

M = % variation in parameter under consideration

T_1 = Increased cycle time

(i) Increased setup cost:

$$C_1 = \left(1+\frac{M}{100}\right)C$$

$$\text{Increase in the cycle time} = T_1 - T$$

$$= \sqrt{\frac{2C_1}{DI(1-D/P)}} - \sqrt{\frac{2C}{DI(1-D/P)}}$$

$$= \sqrt{\frac{2C}{DI(1-D/P)}} - \sqrt{\frac{C_1}{C}} - 1$$

$$= \sqrt{\frac{2C}{DI(1-D/P)}} - \left[\sqrt{\left(1+\frac{M}{100}\right)} - 1\right]$$

And:

$$\%\text{ increase in C.T.} = \sqrt{\left(1+\frac{M}{100}\right)} - 1$$

In order to illustrate, consider the reference parameters as follows:

D	*C*	*I*	*P*	*T*	*E*
600	50	35	960	0.113	887.41

After implementation of the increased setup cost, the approximate effects are given as follows:

% increase in C	5%	10%	15%	20%	25%
C	52.50	55.00	57.50	60.00	62.50
T	0.115	0.118	0.121	0.123	0.126
% increase in T	2.47%	4.88%	7.24%	9.54%	11.80%
E	909.33	930.73	951.64	972.11	992.16
% increase in E	2.47%	4.88%	7.24%	9.54%	11.80%

(ii) Decreased production rate:

$$P_1 = \left(1 - \frac{M}{100}\right)P$$

Increase in the cycle time = $T_1 - T$

$$= \sqrt{\frac{2C}{DI(1-D/P_1)}} - \sqrt{\frac{2C}{DI(1-D/P)}}$$

$$= \sqrt{\frac{2C}{DI(1-D/P)}}\left[\sqrt{\frac{(1-D/P)}{(1-D/P_1)}} - 1\right]$$

$$= \sqrt{\frac{2C}{DI(1-D/P)}}\left[\sqrt{\frac{(1-D/P)}{1-\{D/P(1-M/100)\}}} - 1\right]$$

And:

$$\%\text{ increase in C.T.} = \sqrt{\frac{(1-D/P)}{1-\{D/P(1-M/100)\}}} - 1$$

With the reference parameters as follows:

D	*C*	*I*	*P*	*T*	*E*
600	50	35	960	0.113	887.41

The approximate effects are given below:

% decrease in P	5%	10%	15%	20%	25%	30%
P	912	864	816	768	720	672
T	0.118	0.125	0.134	0.148	0.169	0.211
% Increase in T	4.70%	10.78%	19.02%	30.93%	50.00%	87.08%
E	847.60	801.04	745.58	677.77	591.61	474.34
%decrease in *E*	4.49%	9.73%	15.98%	23.62%	33.33%	46.55%

(iii) Decreased holding cost:

$$I_1 = \left(1 - \frac{M}{100}\right)I$$

Increase in the cycle time = $T_1 - T$

$$= \sqrt{\frac{2C}{DI_1(1-D/P)}} - \sqrt{\frac{2C}{DI(1-D/P)}}$$

$$= \sqrt{\frac{2C}{DI(1-D/P)}}\left[\sqrt{\frac{I}{I_1}} - 1\right]$$

$$= \sqrt{\frac{2C}{DI(1-D/P)}}\left[\sqrt{\frac{1}{(1-M/100)}} - 1\right]$$

And:

$$\% \text{ increase in C.T.} = \sqrt{\frac{1}{(1 - M/100)}} - 1$$

In order to illustrate, consider the reference parameters as before:

D	*C*	*I*	*P*	*T*	*E*
600	50	35	960	0.113	887.41

After implementation of the decreased holding cost, the effects in terms of approximate values are provided below:

% decrease in *I*	5%	10%	15%	20%	25%
I	33.25	31.50	29.75	28.00	26.25
T	0.116	0.119	0.122	0.126	0.130
% increase in *T*	2.60%	5.41%	8.47%	11.80%	15.47%
E	864.94	841.87	818.15	793.73	768.52
% decrease in *E*	2.53%	5.13%	7.80%	10.56%	13.40%

(iv) Increased demand:

$$D_1 = \left(1 + \frac{M}{100}\right) D$$

Increase in the cycle time $= T_1 - T$

$$= \sqrt{\frac{2C}{D_1 I(1 - D_1/P)}} - \sqrt{\frac{2C}{DI(1 - D/P)}}$$

$$= \sqrt{\frac{2C}{DI(1 - D/P)}} \left[\sqrt{\frac{D(1 - D/P)}{D_1(1 - D_1/P)}} - 1 \right]$$

$$= \sqrt{\frac{2C}{DI(1 - D/P)}} \left[\sqrt{\frac{(1 - D/P)}{(1 + M/100)\{1 - D(1 + M/100)/P\}}} - 1 \right]$$

And:

$$\% \text{ increase in C.T.} = \sqrt{\frac{(1 - D/P)}{(1 + M/100)\{1 - D(1 + M/100)/P\}}} - 1$$

The reference parameters are as follows:

D	*C*	*I*	*P*	*T*	*E*
600	50	35	960	0.113	887.41

When demand is increased in this example, the effects in terms of approximate values are given below:

% increase in D	5%	10%	15%	20%	25%
D	630.00	660.00	690.00	720.00	750.00
T	0.115	0.118	0.121	0126	0.132
% increase in T	1.93%	4.45%	7.68%	11.80%	17.11%
E	870.61	849.63	824.15	793.73	757.77
% increase in E	1.89%	4.26%	7.13%	10.56%	14.61%

Table 7.2 summarises the generalised results.

Table 7.2: Results for increased cycle time.

Variation in input parameters	*Increase in optimal C.T.*
Increased facility setup cost	Increase in the cycle time = $\sqrt{\frac{2C}{DI(1-D/P)}}\left[\sqrt{\left(1+\frac{M}{100}\right)}-1\right]$ % increase in C.T. = $\sqrt{\left(1+\frac{M}{100}\right)}-1$
Decreased production rate	Increase in the cycle time = $\sqrt{\frac{2C}{DI(1-D/P)}}\left[\sqrt{\frac{(1\text{-}D/P)}{1-\{D/P(1-M/100)\}}}-1\right]$ % increase in C.T. = $\sqrt{\frac{(1-D/P)}{1-\{D/P(1-M/100)\}}}-1$
Decreased holding cost	Increase in the cycle time = $\sqrt{\frac{2C}{DI(1-D/P)}}\left[\sqrt{\frac{1}{(1-M/100)}}-1\right]$ % increase in C.T. = $\sqrt{\frac{1}{(1-M/100)}}-1$
Increased demand	Increase in the cycle time = $\sqrt{\frac{2C}{DI(1-D/P)}}\left[\sqrt{\frac{(1-D/P)}{(1+M/100)\{1-D(1+M/100)/P\}}}-1\right]$ % increase in C.T. = $\sqrt{\frac{(1-D/P)}{(1+M/100)\{1-D(1+M/100)/P\}}}-1$

7.3 Backordering situation

A cycle of production with shortages that are completely backordered is shown in Fig. 7.5.

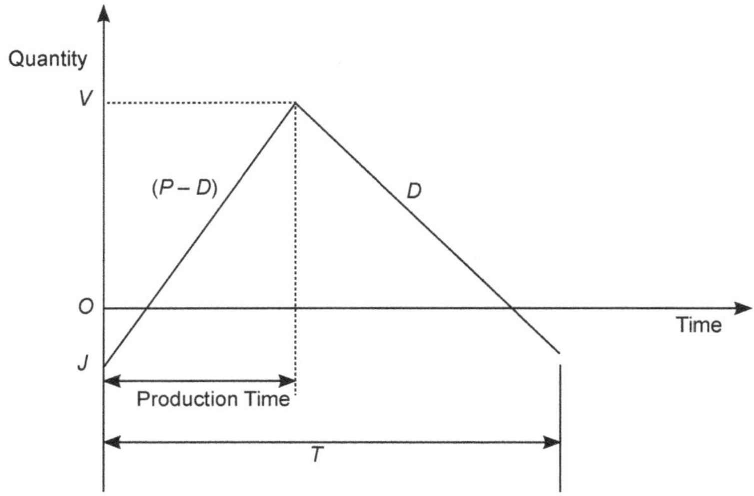

Figure 7.5: A cycle of production with shortages

Maximum shortage quantity = J

$$\text{Period during which shortages occur} = \frac{J}{(P-D)} + \frac{J}{D}$$

$$= \frac{J}{D(1-D/P)}$$

$$\text{Average shortage quantity} = \frac{J}{2}$$

Since there are $\frac{1}{T}$ number of cycles in a year, the annual shortage cost is:

$$ASC = \frac{J}{2} \cdot \frac{J}{D(1-D/P)} \cdot \frac{1}{T} . K$$

where K = Annual shortage cost per unit

Or:

$$ASC = \frac{J^2 K}{2TD(1-D/P)} \tag{7.6}$$

As the production batch size is (DT),

$$\text{Production time} = \frac{DT}{P}$$

However, the production time can also be expressed as:

$$\frac{(V+J)}{(P-D)}$$

Therefore,

$$\frac{(V+J)}{(P-D)} = \frac{DT}{P}$$

Or
$$V = \frac{DT(P-D)}{P} - J$$

Or
$$V = DT(1-D/P) - J \tag{7.7}$$

Positive stock exists during the period:

$$\frac{V}{(P-D)} + \frac{V}{D}$$

$$= \frac{V}{D(1-D/P)}$$

As there are $\frac{1}{T}$ number of cycles in a year, the annual inventory holding cost is given as:

$$AIC = \frac{V}{2}.\frac{V}{D(1-D/P)}.\frac{1}{T}.I$$

Or:

$$AIC = \frac{V^2 I}{2DT(1-D/P)}$$

Substituting the value of V from Eq. (7.7),

$$AIC = \frac{[DT(1-D/P)-J]^2 I}{2DT(1-D/P)}$$

$$= \frac{I[D^2T^2(1-D/P)^2 - 2DTJ(1-D/P) + J^2]}{2DT(1-D/P)}$$

Or:

$$AIC = \frac{IDT(1-D/P)}{2} - IJ + \frac{IJ^2}{2DT(1-D/P)} \tag{7.8}$$

Annual production setup cost is:

$$APC = \frac{C}{T} \tag{7.9}$$

By adding (7.6), (7.8), and (7.9), the total annual cost:

$$E = ASC + AIC + APC$$

$$\text{Or } E = \frac{J^2 K}{2TD(1-D/P)} + \frac{IDT(1-D/P)}{2} - IJ + \frac{IJ^2}{2DT(1-D/P)} + \frac{C}{T} \tag{7.10}$$

Differentiating partially w.r.t. J and equating to zero:

$$\frac{JK}{TD(1-D/P)} - I + \frac{IJ}{DT(1-D/P)} = 0$$

Or $$\frac{J(K+1)}{TD(1-D/P)} = I$$

Or $$J = \frac{IDT(1-D/P)}{(K+I)} \tag{7.11}$$

Substituting the value of J in Eq. (7.10):

$$E = \frac{IDT(1-D/P)}{2} - \frac{I^2DT(1-D/P)}{2(K+1)} + \frac{C}{T} \tag{7.12}$$

In order to obtain the optimal value of T, differentiate with respect to T and equate to zero:

$$\frac{IDT(1-D/P)}{2} - \frac{I^2DT(1-D/P)}{2(K+I)} - \frac{C}{T^2} = 0$$

Or $$\frac{C}{T^2} = \frac{ID(1-D/P)[(K+I)-I]}{2(K+I)}$$

Or $$T^2 = \frac{2C(K+I)}{KID(1-D/P)}$$

Or $$T^* = \sqrt{\frac{2C(K+I)}{KID(1-D/P)}} \tag{7.13}$$

From the Eq. (7.13):

$$\frac{DI(1-D/P)}{2} = \frac{C(K+I)}{KT^2}$$

Putting this value in Eq. (7.12),

$$E^* = T\left[\frac{C(K+I)}{KT^2}\right] - \frac{TI}{(K+I)}\left[\frac{C(K+I)}{KT^2}\right] + \frac{C}{T}$$

Or $$E^* = \frac{C(K+I)}{KT} - \frac{IC}{KT} + \frac{C}{T}$$

Or $$E^* = \frac{C}{T} + \frac{C}{T}$$

Or $$E^* = \frac{2C}{T} \tag{7.14}$$

Example 7.2

Consider the following parameters:

Annual demand, $D = 600$

Annual inventory holding cost per unit, $I =$ ₹ 35

Annual production rate, $P = 960$

Setup cost, $C =$ ₹ 50

Annual shortage cost per unit = ₹ 100

From the Eq. (7.13), cycle time:

$$T = \sqrt{\frac{2 \times 50 \times (100 + 35)}{600 \times 35 \times 100 \times (1 - 600/960)}}$$

Or $$T = 0.1309 \text{ year} \approx 0.131 \text{ year}$$

From the Eq. (7.14), the total cost:

$$E = \frac{2 \times 50}{0.1309}$$

Or $$E = ₹\ 763.76$$

While comparing with the Example 7.1, optimal cycle time is higher in order to incorporate the shortages. However, the overall total cost is lower.

7.3.1 Downward variation in cycle time

Optimal cycle time decreases because of:

(i) Reduced facility setup cost
(ii) Increased production rate
(iii) Increased holding cost
(iv) Decreased demand
(v) Increased shortage cost

Now:

M = % variation in parameter under consideration

T_1 = Reduced cycle time

(i) Reduced setup cost:

Since the parameter under consideration is setup cost, M refers to the % reduction in setup cost.

Now:

$$C_1 = \left(1 - \frac{M}{100}\right) C$$

Decrease in the cycle time $= T - T_1$

$$= \sqrt{\frac{2C(K+I)}{KID(1-D/P)}} - \sqrt{\frac{2C_1(K+I)}{KID(1-D/P)}}$$

$$= \sqrt{\frac{2C(K+I)}{KID(1-D/P)}}\left[1-\sqrt{\frac{C_1}{C}}\right]$$

$$= \sqrt{\frac{2C(K+I)}{KID(1-D/P)}}\left[1-\sqrt{\left(1-\frac{M}{100}\right)}\right]$$

And:

$$\% \text{ decrease in C.T.} = 1-\sqrt{\left(1-\frac{M}{100}\right)}$$

In order to illustrate, consider the parameters of Example 7.2 as follows:

D	*C*	*I*	*P*	*K*	*T*	*E*
600	50	35	960	100	0.131	763.76

After implementation of the reduced setup cost, the approximate effects are given as follows:

% decrease in C	5%	10%	15%	20%	25%	30%
C	47.50	45.00	42.50	40.00	37.50	35.00
T	0.128	0.124	0.121	0.117	0.113	0.110
% decrease in T	2.53%	5.13%	7.80%	10.56%	13.40%	16.33%
E	744.42	724.57	704.15	683.13	661.44	639.01
% decrease in E	2.53%	5.13%	7.80%	10.56%	13.40%	16.33%

(ii) Increased production rate:

$$P_1 = \left(1+\frac{M}{100}\right)P$$

Decrease in the cycle time $= T - T_1$

$$= \sqrt{\frac{2C(K+I)}{KID(1-D/P)}} - \sqrt{\frac{2C(K+I)}{KID(1-D/P_1)}}$$

$$= \sqrt{\frac{2C(K+I)}{KID(1-D/P)}}\left[1-\sqrt{\frac{(1-D/P)}{(1-D/P_1)}}\right]$$

$$= \sqrt{\frac{2C(K+I)}{KID(1-D/P)}}\left[1-\sqrt{\frac{(1-D/P)}{1-\{D/P(1+M/100)\}}}\right]$$

And:

$$\% \text{ decrease in C.T.} = 1 - \sqrt{\frac{(1 - D/P)}{1 - \{D/P(1 + M/100)\}}}$$

In order to illustrate, consider the reference parameters as follows:

D	*C*	*I*	*P*	*K*	*T*	*E*
600	50	35	960	100	0.131	763.76

After implementation of the increased production rate, the effects in terms of approximate values are given as follows:

% increase in P	5%	10%	15%	20%	25%	30%
P	1008	1056	1104	1152	1200	1248
T	0.126	0.122	0.119	0.116	0.113	0.111
% decrease in T	3.75%	6.81%	9.37%	11.53%	13.40%	15.02%
E	793.49	819.58	842.70	863.35	881.92	898.72
% increase in E	3.89%	7.31%	10.34%	13.04%	15.47%	17.67%

% decrease in the cycle time is lower in comparison with the % increase in the production rate.

(iii) Increased holding cost:

$$I_1 = \left(1 + \frac{M}{100}\right) I$$

Decrease in the cycle time $= T - T_1$

$$= \sqrt{\frac{2C(K+I)}{KID(1-D/P)}} - \sqrt{\frac{2C(K+I_1)}{KI_1D(1-D/P)}}$$

$$= \sqrt{\frac{2C(K+I)}{KID(1-D/P)}}\left[1 - \sqrt{\frac{I(K+I_1)}{I_1(K+I)}}\right]$$

$$= \sqrt{\frac{2C(K+I)}{KID(1-D/P)}}\left[1 - \sqrt{\frac{K+I(1+M/100)}{(1+M/100)(K+I)}}\right]$$

And:

$$\% \text{ decrease in C.T.} = 1 - \sqrt{\frac{K+I(1+M/100)}{(1+M/100)(K+I)}}$$

In order to illustrate, consider the reference parameters as before:

D	*C*	*I*	*P*	*K*	*T*	*E*
600	50	35	960	100	0.131	763.76

After implementation of the increased holding cost, the effects in terms of approximate values are provided as follows:

% increase in I	5%	10%	15%	20%	25%	30%
I	36.75	38.50	40.25	42.00	43.75	45.50
T	0.128	0125	0.122	0.120	0.117	0.115
% decrease in T	2.41%	4.65%	6.75%	8.71%	10.56%	12.29%
E	782.62	801.04	819.04	836.66	853.91	870.82
% increase in E	2.47%	4.88%	7.24%	9.54%	11.80%	14.02%

(iv) Decreased demand:

$$D_1 = \left(1-\frac{M}{100}\right)D$$

Decrease in the cycle time $= T - T_1$

$$= \sqrt{\frac{2C(K+I)}{KID(1-D/P)}} - \sqrt{\frac{2C(K+I)}{KID_1(1-D_1/P)}}$$

$$= \sqrt{\frac{2C(K+I)}{KID(1-D/P)}}\left[1-\sqrt{\frac{D(1-D/P)}{D_1(1-D_1/P)}}\right]$$

$$= \sqrt{\frac{2C(K+I)}{KID(1-D/P)}}\left[1-\sqrt{\frac{(1-D/P)}{(1-M/100)\{1-D(1-M/100)/P\}}}\right]$$

And:

$$\%\text{ decrease in C.T.} = 1-\sqrt{\frac{(1-D/P)}{(1-M/100)\{1-D(1-M/100)/P\}}}$$

The reference parameters are as follows:

D	*C*	*I*	*P*	*K*	*T*	*E*
600	50	35	960	100	0.131	763.76

When demand is decreased in this example, the effects in terms of approximate values are given below:

% decrease in D	5%	10%	15%	20%
D	570.00	540.00	510.00	480.00
T	0.129	0.128	0.127	0.127
% decrease in T	1.53	2.29	3.05	3.05
E	774.82	782.62	787.27	788.81
% increase in E	1.45%	2.47%	3.08%	3.28%

(v) Increased shortage cost:

$$K_1 = \left(1+\frac{M}{100}\right)K$$

Decrease in the cycle time = $T - T_1$

$$= \sqrt{\frac{2C(K+I)}{KID(1-D/P)}} - \sqrt{\frac{2C(K_1+I)}{K_1ID(1-D/P)}}$$

$$= \sqrt{\frac{2C(K+I)}{KID(1-D/P)}}\left[1-\sqrt{\frac{K(K_1+I)}{K_1(K+I)}}\right]$$

$$= \sqrt{\frac{2C(K+I)}{KID(1-D/P)}}\left[1-\sqrt{\frac{I+K(1+M/100)}{(1+M/100)(K+I)}}\right]$$

And:

$$\% \text{ decrease in C.T.} = 1-\sqrt{\frac{I+K(1+M/100)}{(1+M/100)(K+I)}}$$

With the reference parameters as follows:

D	*C*	*I*	*P*	*K*	*T*	*E*
600	50	35	960	100	0.131	763.76

The effects in terms of approximate values are given below:

% increase in K	5%	10%	15%	20%	25%	30%
K	105	110	115	120	125	130
T	0.1301	0.1294	0.1287	0.1281	0.1275	0.1270
% decrease in T	0.62%	1.19%	1.71%	2.18%	2.63%	3.04%
E	768.52	772.93	777.01	780.82	784.37	787.69
% increase in E	0.62%	1.20%	1.73%	2.23%	2.70%	3.13%

Table 7.3 summarises the generalised results.

Table 7.3: Results for reduced cycle time with shortages.

Variation in input parameter	*Decrease in optimal C.T.*
Reduced facility setup cost	Decrease in the cycle time $= \sqrt{\frac{2C(K+I)}{KID(1-D/P)}}\left[1-\sqrt{\left(1-\frac{M}{100}\right)}\right]$ % decrease in C.T. = $1-\sqrt{\left(1-\frac{M}{100}\right)}$

Contd...

Contd...

Variation in input parameter	*Decrease in optimal C.T.*
Increased production rate	Decrease in the cycle time $= \sqrt{\frac{2C(K+I)}{KID(1-D/P)}}\left[1-\sqrt{\frac{(1-D/P)}{1-\{D/P(1+M/100)\}}}\right]$ % decrease in C.T. $= 1-\sqrt{\frac{(1-D/P)}{1-\{D/P(1+M/100)\}}}$
Increased holding cost	Decrease in the cycle time $= \sqrt{\frac{2C(K+I)}{KID(1-D/P)}}\left[1-\sqrt{\frac{K+I(1+M/100)}{(1+M/100)(K+I)}}\right]$ % decrease in C.T. $= 1-\sqrt{\frac{K+I(1+M/100)}{(1+M/100)(K+I)}}$
Decreased demand	Decrease in the cycle time $= \sqrt{\frac{2C(K+I)}{KID(1-D/P)}}\left[1-\sqrt{\frac{(1-D/P)}{(1-M/100)\{1-D(1-M/100)/P\}}}\right]$ % decrease in C.T. $= 1-\sqrt{\frac{(1-D/P)}{(1-M/100)\{1-D(1-M/100)/P\}}}$
Increased shortage cost	Decrease in the cycle time $= \sqrt{\frac{2C(K+I)}{KID(1-D/P)}}\left[1-\sqrt{\frac{I+K(1+M/100)}{(1+M/100)(K+I)}}\right]$ % decrease in C.T. $= 1-\sqrt{\frac{I+K(1+M/100)}{(1+M/100)(K+I)}}$

7.3.2 Upward variation in cycle time

Optimal cycle time increases because of:

(i) Increased facility setup cost
(ii) Reduced production rate
(iii) Reduced holding cost
(iv) Increased demand
(v) Reduced shortage cost

Now:

M = % variation in parameter under consideration

T_1 = Increased cycle time

(i) Increased setup cost:

Since the parameter under consideration is setup cost, M refers to the % increase in setup cost.

Now:

$$C_1 = \left(1+\frac{M}{100}\right)C$$

Increase in the cycle time = $T_1 - T$

$$= \sqrt{\frac{2C_1\ (K+I)}{KID(1-D/P)}} - \sqrt{\frac{2C(K+I)}{KID(1-D/P)}}$$

$$= \sqrt{\frac{2C(K+I)}{KID(1-D/P)}}\left[\sqrt{\frac{C_1}{C}}-1\right]$$

$$= \sqrt{\frac{2C(K+I)}{KID(1-D/P)}}\left[\sqrt{\left(1+\frac{M}{100}\right)}-1\right]$$

And: % increase in C.T. = $\sqrt{\left(1+\frac{M}{100}\right)}-1$

In order to illustrate, consider the parameters of Example 7.2 as follows:

D	*C*	*I*	*P*	*K*	*T*	*E*
600	50	35	960	100	0.131	763.76

After implementation of the increased setup cost, the effects are given as follows:

% increase in C	5%	10%	15%	20%	25%	30%
C	52..50	55.00	57.50	60.00	62.50	65.00
T	0.134	0.137	0.140	0.143	0.146	0.149
% increase in T	2.47%	4.88%	7.24%	9.54%	11.80%	14.02%
E	782.62	801.04	819.04	836.66	853.91	870.82
% increase in E	2.47%	4.88%	7.24%	9.54%	11.80%	14.02%

(ii) Reduced production rate:

$$P_1 = \left(1-\frac{M}{100}\right)P$$

Increase in the cycle time $= T_1 - T$

$$= \sqrt{\frac{2C(K+I)}{KID(1-D/P_1)}} - \sqrt{\frac{2C(K+I)}{KID(1-D/P)}}$$

$$= \sqrt{\frac{2C(K+I)}{KID(1-D/P)}}\left[\sqrt{\frac{(1-D/P)}{(1-D/P_1)}}-1\right]$$

$$= \sqrt{\frac{2C(K+I)}{KID(1-D/P)}}\left[\sqrt{\frac{(1-D/P)}{1-\{D/P(1-M/100)\}}}-1\right]$$

And:

$$\% \text{ increase in C.T.} = \sqrt{\frac{(1-D/P)}{1-\{D/P(1-M/100)\}}}-1$$

In order to illustrate, consider the reference parameters as follows:

D	*C*	*I*	*P*	*K*	*T*	*E*
600	50	35	960	100	0.131	763.76

After implementation of the reduced production rate, the effects in terms of approximate values are given as follows:

% decrease in P	5%	10%	15%	20%	25%	30%
P	912	864	816	765	720	672
T	0.137	0.145	0.156	0.171	0.196	0.245
% increase in T	4.70%	10.78%	19.02%	30.93%	50.00%	87.08%
E	729.50	689.43	641.69	583.33	509.18	408.25
% decrease in E	4.49%	9.73%	15.98%	23.62%	33.33%	46.55%

(iii) Reduced holding cost:

$$I_1 = \left(1-\frac{M}{100}\right)I$$

Increase in the cycle time $= T_1 - T$

$$= \sqrt{\frac{2C(K+I_1)}{KI_1D(1-D/P)}} - \sqrt{\frac{2C(K+I)}{KID(1-D/P)}}$$

$$= \sqrt{\frac{2C(K+I)}{KID(1-D/P)}}\left[\sqrt{\frac{I(K+I_1)}{I_1(K+1)}}-1\right]$$

$$= \sqrt{\frac{2C(K+I)}{KID(1-D/P)}}\left[\sqrt{\frac{K+I(1-M/100)}{1-M/100)(K+I)}}-1\right]$$

And:

$$\% \text{ increase in C.T.} = \sqrt{\frac{K+I(1-M/100)}{1-M/100)(K+I)}}-1$$

In order to illustrate, consider the reference parameters as before:

D	*C*	*I*	*P*	*K*	*T*	*E*
600	50	35	960	100	0.131	763.76

After Implementation of the reduced holding cost, the effects in terms of approximate values are provided below:

%decrease in I	5%	10%	155	20%	25%	30%
I	33.25	31.50	29.75	28.00	26.25	24.50
T	0.134	0.138	0.142	0.146	0.151	0.156
% increase in T	2.60%	5.41%	8.47%	11.80%	15.47%	19.52%
E	744.42	724.57	704.15	683.13	661.44	639.01
% decrease in E	2.53%	5.13%	7.80%	10.56%	13.40%	16.33%

(iv) Increased demand:

$$D_1 = \left(1+\frac{M}{100}\right)D$$

Increase in the cycle time $= T_1 - T$

$$= \sqrt{\frac{2C(K+I)}{KID_1(1-D_1/P)}} - \sqrt{\frac{2C(K+I)}{KID(1-D/P)}}$$

$$= \sqrt{\frac{2C(K+I)}{KID(1-D/P)}}\left[\sqrt{\frac{D(1-D/P)}{D_1(1-D_1/P)}}-1\right]$$

$$= \sqrt{\frac{2C(K+I)}{KID(1-D/P)}}\left[\sqrt{\frac{(1-D/P)}{(1+M/100)\{1-D(1+M/100)/P\}}}-1\right]$$

And:

$$\% \text{ increase in C.T.} = \sqrt{\frac{(1-D/P)}{(1+M/100)\{1-D(1+M/100)/P\}}}-1$$

The reference parameters are as follows:

D	*C*	*I*	*P*	*K*	*T*	*E*
600	50	35	960	100	0.131	763.73

When demand is increased in this example, the effects in terms of approximate values are given below:

% increase in D	5%	10%	15%	20%	25%	30.00%
D	630.00	660.00	690.00	720.00	750.00	780.00
T	0.133	0.137	0.141	0.146	0.153	0.162
% increase in T	1.93%	4.45%	7.68%	11.80%	17.11%	24.03%
E	749.31	731.25	709.31	683.13	652.19	615.77
% decrease in E	1.89%	4.26%	7.13%	10.56%	14.61%	19.38%

(v) Reduced shortage cost:

$$K_1 = \left(1-\frac{M}{100}\right)K$$

Increase in the cycle time $= T_1 - T$

$$= \sqrt{\frac{2C(K_1+I)}{K_1ID(1-D/P)}} - \sqrt{\frac{2C(K+I)}{KID(1-D/P)}}$$

$$= \sqrt{\frac{2C(K+I)}{KID(1-D/P)}}\left[\sqrt{\frac{K(K_1+I)}{K_1(K+I)}}-1\right]$$

$$= \sqrt{\frac{2C(K+I)}{KID(1-D/P)}}\left[\sqrt{\frac{I+K(1-M/100)}{(1-M/100)(K+I)}}-1\right]$$

And:

$$\% \text{ increase in C.T.} = \sqrt{\frac{I+K(1-M/100)}{(1-M/100)(K+I)}}-1$$

With the reference parameters as follows:

D	*C*	*I*	*P*	*K*	*T*	*E*
600	50	35	960	100	0.131	763.76

The effects in terms of approximate values are given below:

% decrease in K	5%	10%	15%	20%	25%	30%
K	95	90	85	85	75	70
T	0.1318	0.1328	0.1339	0.1351	0.1365	0.1380
% increase in T	0.68%	1.43%	2.26%	3.19%	4.23%	5.41%
E	758.60	752.99	746.87	740.15	732.76	724.57
% decrease in E	0.68%	1.41%	2.21%	3.09%	4.06%	5.13%

Table 7.4 summarises the generalised results.

Table 7.4: Results for increased cycle time with shortages.

Variation in input parameter	*Increase in optimal C.T.*
Increased facility setup cost	Increase in the cycle time $= \sqrt{\frac{2C(K+I)}{KID(1-D/P)}}\left[\sqrt{\left(1+\frac{M}{100}\right)}-1\right]$ % Increase in C.T. $= \sqrt{\left(1+\frac{M}{100}\right)}-1$
Reduced production rate	Increase in the cycle time $= \sqrt{\frac{2C(K+I)}{KID(1-D/P)}}\left[\sqrt{\frac{(1-D/P)}{1-\{D/P(1-M/100)\}}}-1\right]$ % increase in C.T. $= \sqrt{\frac{(1-D/P)}{1-\{D/P(1-M/100)\}}}-1$
Reduced holding cost	Increase in the cycle time $= \sqrt{\frac{2C(K+I)}{KID(1-D/P)}}\left[\sqrt{\frac{K+I(1-M/100)}{(1-M/100)(K+I)}}-1\right]$ % Increase in C.T. $= \sqrt{\frac{K+I(1-M/100)}{(1-M/100)(K+I)}}-1$
Increased demand	Increase in the cycle time $= \sqrt{\frac{2C(K+I)}{KID(1-D/P)}}\left[\sqrt{\frac{(1-D/P)}{(1+M/100)\{1-D(1+M/100)/P\}}}-1\right]$ % Increase in C.T. $= \sqrt{\frac{(1-D/P)}{(1+M/100)\{1-D(1+M/100)/P\}}}-1$
Reduced shortage cost	Increase in the cycle time $= \sqrt{\frac{2C(K+I)}{KID(1-D/P)}}\left[\sqrt{\frac{I+K\,(1-M/100)}{(1-M/100)(K+I)}}-1\right]$ % Increase in C.T. $= \sqrt{\frac{I+K\,(1-M/100)}{(1-M/100)(K+I)}}-1$

7.4 Interaction of parameters without shortages

When shortages are not allowed, production cycle time is determined on the basis of optimisation related to the remaining various parameters. However, when a parameter varies, the cycle time also changes. In many situations, either because of convenience in material handling or certain arrangement between buyer and supplier, a similar cycle time is preferred. Suitable management response should be available by way of change in another parameter.

7.4.1 Increased cycle time

Cycle time increases because of:

(a) Higher setup cost
(b) Lower production rate
(c) Lower holding cost
(d) Higher demand

(a) Higher setup cost:

Consider the following parameters:

Annual demand, $D = 600$

Annual production holding cost per unit, $I = ₹\ 35$

Annual production rate, $P = 960$

Setup cost, $C = ₹\ 50$

From the Eq. (7.4), optimal cycle time:

$$T = \sqrt{\frac{2C}{DI(1 - D/P)}}$$

Or

$$T = \sqrt{\frac{2 \times 50}{600 \times 35(1 - 600/960)}}$$

Or

$$T = 0.1127 \text{ year}$$

Now if setup cost is increased by 10% then

$$C_1 = ₹\ 55$$

And the corresponding cycle time will increase to 0.1182 year. However, if an objective is to have a similar cycle time as before, then a response might be in the form of production rate increase. The increased production rate can be obtained as follows:

$$\sqrt{\frac{2\times 55}{600\times 35(1-600/P_1)}} = 0.1127$$

Or $$P_1 = 1021.11$$

That is, approximately 6.37% increase in production rate.

For a general approach, let:

M = % variation in the parameter that triggers the change

N = % variation in the response parameter in order to have similar cycle time

Now:

$$C_1 = C\left(1+\frac{M}{100}\right)$$

$$P_1 = P\left(1+\frac{N}{100}\right)$$

For similar cycle time:

$$\sqrt{\frac{2C}{DI(1-D/P)}} = \sqrt{\frac{2C_1}{DI(1-D/P_1)}}$$

Or $$C(1-D/P_1) = C_1(1-D/P)$$

Or $$1-\frac{D}{P_1} = 1-(1-M/100)(1-D/P)$$

Or $$\frac{D}{P_1} = 1-(1+M/100)(1-D/P)$$

Or $$P_1 = \frac{D}{1-(1+M/100)(1-D/P)}$$

Or $$1+\frac{N}{100} = \frac{(D/P)}{1-(1+M/100)(1-D/P)}$$

Or $$\frac{N}{100} = \frac{(D/P)-1+(1+M/100)(1-D/P)}{1-(1+M/100)(1-D/P)}$$

Or $$\frac{N}{100} = \frac{(D/P)-1+1-(D/P)+(M/100)-(M/100)(D/P)}{1-[(1-(D/P)+(M/100)-(M/100)(D/P)]}$$

Or $$\frac{N}{100} = \frac{(M/100)(1-D/P)}{(D/P)-(M/100)+(M/100)(D/P)}$$

Or $$N = \frac{M(1-D/P)}{(D/P)(1+M/100)-(M/100)}$$

For the stated reference data, variation of N with respect to M is shown as follows:

S. No.	M	$N = \frac{M(1 - D/P)}{(D/P)(1 + M/100) - (M/100)}$
1	5	3.09
2	10	6.38
3	15	9.89
4	20	13.64
5	25	17.65

The values of N are lower than that of M. However, these are more sensitive towards higher values of M.

(b) Lower Production rate:

Reduction in the setup cost might be an option for similar cycle time.

$$P_1 = P\left(1 - \frac{M}{100}\right)$$

$$C_1 = C\left(1 - \frac{N}{100}\right)$$

Now:

$$\sqrt{\frac{2C}{DI(1 - D/P)}} = \sqrt{\frac{2C_1}{DI(1 - D/P_1)}}$$

Or $$C_1(1 - D/P) = C(1 - D/P_1)$$

Or $$1 - \frac{N}{100} = \frac{(1 - D/P_1)}{(1 - D/P)}$$

Or $$\frac{N}{100} = \frac{(1 - D/P) - (1 - D/P_1)}{(1 - D/P)}$$

Or $$\frac{N}{100} = \frac{(D/P_1) - (D/P)}{(1 - D/P)}$$

Or $$\frac{N}{100} = \frac{D - D(P_1/P)}{P_1(1 - D/P)}$$

Or $$\frac{N}{100} = \frac{D[1 - 1 + (M/100)]}{P(1 - M/100)(1 - D/P)}$$

Or $$\frac{N}{100} = \frac{(D/P)(M/100)}{(1 - M/100)(1 - D/P)}$$

Or $$N = \frac{M(D/P)}{(1-M/100)(1-D/P)}$$

With the reference parameters as before:

D	*C*	*I*	*P*	*T*	*E*
600	50	35	960	0.113	887.41

And with the use of relevant parameters, the variation of N is shown below:

S. No.	*M*	$N = \frac{M(D/P)}{(1-M/100)(1-D/P)}$
1	2	3.40
2	4	6.94
3	6	10.64
4	8	14.49
5	10	18.52

The values of N are higher than that of M and also are more sensitive towards higher values of M.

(c) Lower holding cost:

Increased production rate might be one of the options for similar cycle time.

$$I_1 = I\left(1 - \frac{M}{100}\right)$$

$$P_1 = P\left(1 + \frac{N}{100}\right)$$

$$\sqrt{\frac{2C}{DI(1-D/P)}} = \sqrt{\frac{2C}{DI_1(1-D/P_1)}}$$

Or $$I_1(1-D/P_1) = I(1-D/P)$$

Or $$1 - \frac{D}{P_1} = \frac{(1-D/P)}{(1-M/100)}$$

Or $$\frac{D}{P_1} = \frac{1-(M/100)-1+(D/P)}{(1-M/100)}$$

Or $$P_1 = \frac{D(1-M/100)}{(D/P)-(M/100)}$$

Or $$1 + \frac{N}{100} = \frac{(D/P)(1-M/100)}{(D/P)-(M/100)}$$

Or $$\frac{N}{100} = \frac{(D/P)(1-M/100)-(D/P)+(M/100)}{(D/P)-(M/100)}$$

Or $$\frac{N}{100} = \frac{(M/100)-(D/P)(M/100)}{(D/P)-(M/100)}$$

Or $$\frac{N}{100} = \frac{(M/100)(1-D/P)}{(D/P-M/100)}$$

Or $$N = \frac{M(1-D/P)}{(D/P)-(M/100)}$$

With the reference parameters including D and P as 600 and 960, respectively, the variation of N is provided below:

S. No.	M	$N = \frac{M(1-D/P)}{(D/P)-(M/100)}$
1	2	1.24
2	4	2.56
3	6	3.98
4	8	5.50
5	10	7.14

The values of N are lower than that of M, but these are more sensitive towards higher values of M.

Another option can be a setup cost reduction.

$$I_1 = I\left(1-\frac{M}{100}\right)$$

$$C_1 = C\left(1-\frac{N}{100}\right)$$

$$\sqrt{\frac{2C}{DI(1-D/P)}} = \sqrt{\frac{2C_1}{DI_1(1-D/P)}}$$

Or $$C_1 I = CI_1$$

Or $$1-\frac{N}{100} = 1-\frac{M}{100}$$

Or $$N = M$$

(d) Higher demand

Setup cost reduction and higher production rate are considered as the options.

(i) Setup cost reduction

$$D_1 = D\left(1+\frac{M}{100}\right)$$

$$C_1 = C\left(1-\frac{N}{100}\right)$$

Now:

$$\sqrt{\frac{2C}{DI(1-D/P)}} = \sqrt{\frac{2C_1}{D_1 I(1-D_1/P)}}$$

Or $C_1 D(1-D/P) = C_1 D_1(1-D_1/P)$

Or $C_1 = \dfrac{C(1+M/100)(1-D_1/P)}{(1-D/P)}$

Or $1-\dfrac{N}{100} = \dfrac{(1+M/100)(1-D_1/P)}{(1-D/P)}$

Or $\dfrac{N}{100} = \dfrac{(1-D/P)-(1+M/100)(1-D_1/P)}{(1-D/P)}$

Or $\dfrac{P_1}{D_1} = \dfrac{1-(D/P)-(1+M/100)+(1+M/100)(D_1/P)}{(1-D/P)}$

Or $\dfrac{N}{100} = \dfrac{(1+M/100)^2\ (D/P)-(D/P)-(M/100)}{(1-D/P)}$

Or $\dfrac{N}{100} = \dfrac{(D/P)[1+(2M/100)+(M/100)^2-1]-(M/100)}{(1-D/P)}$

Or $N = \dfrac{(D/P)[2M+(M^2/100)]-M}{(1-D/P)}$

With the reference parameters:

D	*C*	*I*	*P*	*T*	*E*
600	50	35	960	0.113	887.41

And with the use of relevant parameters, the variation of *N* is shown below:

S. No.	*M*	$N = \dfrac{(D/P)[2M+(M^2/100)]-M}{(1-D/P)}$
1	2	1.40
2	4	2.93
3	6	4.60
4	8	6.40
5	10	8.33

The values of N are lower than that of M, but these are more sensitive towards higher values of M.

(ii) Higher production rate

$$D_1 = D\left(1 + \frac{M}{100}\right)$$

$$P_1 = P\left(1 + \frac{N}{100}\right)$$

$$\sqrt{\frac{2C}{DI(1 - D/P)}} = \sqrt{\frac{2C}{D_1 I(1 - D_1/P)}}$$

Or $$D_1(1 - D_1/P_1) = D(1 - D/P)$$

Or $$1 - \frac{D_1}{P_1} = \frac{(1 - D/P)}{(1 + M/100)}$$

Or $$\frac{D_1}{P_1} = \frac{1 + (M/100) - 1 + (D/P)}{(1 + M/100)}$$

Or $$\frac{P_1}{D_1} = \frac{(1 + M/100)}{(M/100) + (D/P)}$$

Or $$P\left(1 + \frac{N}{100}\right) = \frac{D(1 + M/100)^2}{(M/100) + (D/P)}$$

O $$1 + \frac{N}{100} = \frac{(D/P)(1 + M/100)^2}{(M/100) + (D/P)}$$

Or $$\frac{N}{100} = \frac{(D/P)(1 + M/100)^2 - (M/100) - (D/P)}{(M/100) + (D/P)}$$

Or $$\frac{N}{100} = \frac{(D/P)[1 + (2M/100) + (M/100)^2 - 1] - (M/100)}{(M/100) + (D/P)}$$

Or $$N = \frac{(D/P)[2M + (M^2/100)] - M}{(M/100) + (D/P)}$$

With the reference parameters as before, the variation of N is shown as follows:

S. No.	M	$N = \frac{(D/P)[2M + (M^2/100)] - M}{(M/100) + (D/P)}$
1	2	0.81
2	4	1.65
3	6	2.52
4	8	3.40
5	10	4.31

The values of *N* are much lower than that of *M*, but these are slightly more sensitive towards higher values of *M*. The *N* values are lower in comparison with the previous option. However, ease in implementation (that varies from organization to organization) should also be considered before finally selecting a suitable alternative.

7.4.2 Decreased cycle time

Cycle time decreases because of:

(a) Lower setup cost
(b) Higher production rate
(c) Higher holding cost
(d) Lower demand

(a) Lower setup cost:

Production rate decrease might be an option.

$$C_1 = C\left(1 - \frac{M}{100}\right)$$

$$P_1 = P\left(1 - \frac{N}{100}\right)$$

For similar cycle time:

$$\sqrt{\frac{2C}{DI(1 - D/P)}} = \sqrt{\frac{2C_1}{DI(1 - D/P_1)}}$$

Or $(1 - D/P_1)C = (1 - D/P)C_1$

Or $1 - \frac{D}{P_1} = (1 - D/P)(1 - M/100)$

O $\frac{D}{P_1} = 1 - (1 - D/P)(1 - M/100)$

Or $$P_1 = \frac{D}{1-(1-D/P)(1-M/100)}$$

Or $$1-\frac{N}{100} = \frac{(D/P)}{1-(1-D/P)(1-M/100)}$$

Or $$\frac{N}{100} = \frac{1-(1-D/P)(1-M/100)-(D/P)}{1-(1-D/P)(1-M/100)}$$

Or $$\frac{N}{100} = \frac{(1-D/P)(1-1-M/100)}{1-(1-D/P)(1-M/100)}$$

Or $$N = \frac{M(1-D/P)}{1-(1-D/P)(1-M/100)}$$

For the reference data:

D	*C*	*I*	*P*	*T*	*E*
600	50	35	960	0.113	887.41

Variation of *N* with respect to *M* is shown as follows:

S. No.	*M*	$N = \frac{M(1-D/P)}{1-(1-D/P)(1-M/100)}$
1	5	2.91
2	10	5.66
3	15	8.26
4	20	10.71
5	25	13.04

The values of *N* are lower than that of *M* and are also less sensitive towards higher values of *M*.

(b) Higher production rate:

Holding cost reduction might be an option.

$$P_1 = P\left(1+\frac{M}{100}\right)$$

$$I_1 = I\left(1-\frac{N}{100}\right)$$

Now:

$$\sqrt{\frac{2C}{DI(1-D/P)}} = \sqrt{\frac{2C}{DI_1(1-D/P_1)}}$$

Or $\quad I_1(1 - D/P_1) = I(1 - D/P)$

Or $\quad 1 - \dfrac{N}{100} = \dfrac{(1 - D/P)}{(1 - D/P_1)}$

Or $\quad \dfrac{N}{100} = \dfrac{(1 - D/P_1) - (1 - D/P)}{(1 - D/P_1)}$

Or $\quad \dfrac{N}{100} = \dfrac{(D/P) - (D/P_1)}{(1 - D/P_1)}$

Or $\quad \dfrac{N}{100} = \dfrac{D(P_1/P) - D}{(P_1 - D)}$

Or $\quad \dfrac{N}{100} = \dfrac{D[(1 + M/100) - 1)]}{P(1 + M/100) - D}$

Or $\quad \dfrac{N}{100} = \dfrac{(DM/100)}{(PM/100) + (P - D)}$

Or $\quad N = \dfrac{DM}{(PM/100) + (P - D)}$

With the reference parameters including D and P as 600 and 960, respectively, the variation of N is provided below:

S. No.	M	$N = \dfrac{DM}{(PM/100) + (P - D)}$
1	2	3.16
2	4	6.02
3	6	8.52
4	8	10.99
5	10	13.16

The values of N are higher than that of M, but these are less sensitive towards greater values of M.

(c) Higher holding cost:

Consider a production rate decrease for the management response.

$$I_1 = I\left(1 + \frac{M}{100}\right)$$

$$P_1 = P\left(1 - \frac{N}{100}\right)$$

$$\sqrt{\frac{2C}{DI(1 - D/P)}} = \sqrt{\frac{2C}{DI_1(1 - D/P_1)}}$$

Or $I_1(1 - D/P_1) = I(1 - D/P)$

Or $1 - \dfrac{D}{P_1} = \dfrac{(1 - D/P)}{(1 + M/100)}$

Or $\dfrac{D}{P_1} = \dfrac{(1 + M/100) - 1 + (D/P)}{(1 + M/100)}$

Or $\dfrac{P_1}{D} = \dfrac{(1 + M/100)}{(M/100) + (D/P)}$

Or $1 - \dfrac{N}{100} = \dfrac{(D/P)(1 + M/100)}{(M/100) + (D/P)}$

Or $\dfrac{N}{100} = \dfrac{(M/100) + (D/P) - (D/P) - (D/P)(M/100)}{(M/100) + (D/P)}$

Or $\dfrac{N}{100} = \dfrac{(M/100)(1 - D/P)}{(M/100) + (D/P)}$

Or $N = \dfrac{M(1 - D/P)}{(M/100) + (D/P)}$

For the reference data:

D	C	I	P	T	E
600	50	35	960	0.113	887.41

Variation of N with respect to M is shown as follows:

S. No.	M	$N = \dfrac{M(1-D/P)}{(M/100)+(D/P)}$
1	5	2.78
2	10	5.17
3	15	7.26
4	20	9.09
5	25	10.71

The values of N are lower than that of M and are also less sensitive towards higher of M.

(d) Lower demand:

Consider a reduction in holding cost for similar cycle time.

$$D_1 = D\left(1 - \frac{M}{100}\right)$$

$$I_1 = I\left(1 - \frac{N}{100}\right)$$

$$\sqrt{\frac{2C}{DI(1-D/P)}} = \sqrt{\frac{2C}{D_1 I_1(1-D_1/P)}}$$

Or $\quad D_1 I_1(1 - D_1/P) = DI(1 - D/P)$

Or $\quad (1 - M/100)(1 - N/100) = \dfrac{(1-D/P)}{(1-D_1/P)}$

Or $\quad 1 - \dfrac{N}{100} = \dfrac{(1-D/P)}{(1-M/100)(1-D_1/P)}$

Or $\quad \dfrac{N}{100} = \dfrac{(1-M/100)(1-D_1/P)-(1-D/P)}{(1-M/100)(1-D_1/P)}$

Or $\quad \dfrac{N}{100} = \dfrac{1-(D_1/P)-(M/100)+(M/100)(D_1/P)-1+(D/P)}{(1-M/100)\{(1-(D/P)(1-M/100)\}}$

Or $\quad \dfrac{N}{100} = \dfrac{(D/P)-(D_1/P)-(M/100)(1-D_1/P)}{(1-M/100)\{(1-D/P)(1-M/100)\}}$

Or $\quad \dfrac{N}{100} = \dfrac{(D/P)-(D/P)(1-M/100)-(M/100)(1-D_1/P)}{(1-M/100)\{1-(D/P)(1-M/100)\}}$

Or $\quad \dfrac{N}{100} = \dfrac{(D/P)(1-1+M/100)-(M/100)\{1-(D/P)(1-M/100)\}}{(1-M/100)\{1-(D/P)(1-M/100)\}}$

Or $\quad \dfrac{N}{100} = \dfrac{(D/P)(M/100)-(M/100)\{1-(D/P)(1-M/100)\}}{(1-M/100)\{1-(D/P)(1-M/100)\}}$

Or $\quad N = \dfrac{M(D/P)-M\{1-(D/P)(1-M/100)\}}{(1-M/100)\{1-(D/P)(1-M/100)\}}$

With the reference parameters including D and P as 600 and 960, respectively, the variation of N is provided below:

S. No.	M	$N = \dfrac{M(D/P)-M\{1-(D/P)(1-M/100)\}}{(1-M/100)\{1-(D/P)(1-M/100)\}}$
1	2	1.25
2	4	2.34
3	6	3.29
4	8	4.09
5	10	4.76

The values of N are lower than that of M and also are less sensitive towards greater values of M.

7.5 Interaction of parameters with shortages

With the inclusion of shortages, the interaction is analysed concerning various parameters.

7.5.1 Longer cycle time

Cycle time increases also because of a reduction in shortage cost. For a similar cycle time, consider the management response in the form of:

(i) Reduced setup cost
(ii) Increased production rate

For a general approach, let:

M = % variation in the shortage cost

N = % variation in the response parameter

(i) Reduced setup cost:

$$K_1 = K\left(1 - \frac{M}{100}\right)$$

$$C_1 = C\left(1 - \frac{N}{100}\right)$$

Since the optimal cycle time is given as:

$$T = \sqrt{\frac{2C(K+I)}{KID(1-D/P)}}$$

For similar cycle time:

$$\sqrt{\frac{2C(K+I)}{KID(1-D/P)}} = \sqrt{\frac{2C_1(K_1+I)}{K_1ID(1-D/P)}}$$

Or $$\frac{C(K+I)}{K} = \frac{C_1(K_1+I)}{K_1}$$

Or $$C_1 = \frac{K_1C(K+I)}{K(K_1+I)}$$

Or $$1 - \frac{N}{100} = \frac{(1-M/100)(K+I)}{I+K(1-M/100)}$$

Or $$\frac{N}{100} = \frac{I + K(1 - M/100) - (1 - M/100)(K + I)}{I + K(1 - M/100)}$$

Or $$\frac{N}{100} = \frac{I - I(1 - M/100)}{I + K(1 - M/100)}$$

Or $$\frac{N}{100} = \frac{(IM/100)}{I + K(1 - M/100)}$$

Or $$N = \frac{IM}{I + K(1 - M/100)}$$

For the reference parameters as follows:

D	*C*	*I*	*P*	*K*	*T*	*E*
600	50	35	960	100	0.131	763.76

The values of *N* are given w.r.t as follows:

S. No.	*M*	$N = \frac{IM}{I + K(1 - M/100)}$
1	5	1.35
2	10	2.80
3	15	4.38
4	20	6.09
5	25	7.95

The values of *N* are much lower than that of *M* but are more sensitive towards greater values of *M*.

(ii) Increased production rate:

$$K_1 = K\left(1 - \frac{M}{100}\right)$$

$$P_1 = P\left(1 + \frac{N}{100}\right)$$

For a similar cycle time:

$$\sqrt{\frac{2C(K + I)}{KID(1 - D/P)}} = \sqrt{\frac{2C(K_1 + I)}{K_1 ID(1 - D/P_1)}}$$

Or $$\frac{(K + I)}{K(1 - D/P)} = \frac{(K_1 + I)}{K_1(1 - D/P_1)}$$

With the use of the above expression and following the process explained before, it can be analysed furthermore. However, in certain cases, a combination of parameters may also be useful. For example, with reference to the previous option (i.e., the reduced setup cost):

$$M = 15; N = 4.38$$

However, in certain operational setting, it the management feels that the value of N cannot be more then 3 (say), then a combination of parameters may be considered.

Combination of setup cost and production rate:

Let M, N, and R be the % variation in shortage cost, setup cost, and production rate.

Now:

$$K_1 = K\left(1 - \frac{M}{100}\right)$$

$$C_1 = C\left(1 - \frac{N}{100}\right)$$

$$P_1 = P\left(1 + \frac{R}{100}\right)$$

For a similar cycle time:

$$\sqrt{\frac{2C(K+I)}{KID(1-D/P)}} = \sqrt{\frac{2C_1(K+I)}{K_1ID(1-D/P_1)}}$$

Or $$\frac{C(K+I)}{K(1-D/P)} = \frac{C_1(K_1+I)}{K_1(1-D/P_1)}$$

Or $$1 - \frac{D}{P_1} = \frac{KC_1(K_1+I)(1-D/P)}{K_1C(K+I)}$$

Or $$\frac{D}{P_1} = \frac{K_1C(K+I) - KC_1(K_1+I)(1-D/P)}{K_1C\ (K+I)}$$

Or $$\frac{P_1}{D} = \frac{K_1C(K+I\ )}{K_1C(K+I) - KC_1(K_1+I)(1-D/P)}$$

Or $$1 + \frac{R}{100} = \frac{(D/P)K_1C(K+I\ )}{K_1C(K+I) - KC_1(K_1+I)(1-D/P)}$$

Or $$\frac{R}{100} = \frac{(D/P)K_1C(K+I\ ) - K_1C(K+I) + KC_1(K_1+I)(1-D/P)}{K_1C(K+I) - KC_1\ (K_1+I)(1-D/P)}$$

Or $$\frac{R}{100} = \frac{KC_1(K_1+I)(1-D/P) - K_1C(K+I)(1-D/P)}{K_1C(K+I) - KC_1\ (K_1+I)(1-D/P)}$$

Or $$\frac{R}{100} = \frac{(1-D/P)[KC_1(K_1+I)-K_1C(K+I)]}{KC[(1-M/100)(K+I)-(1-N/100)(K_1+I)(1-D/P)]}$$

Or $$\frac{R}{100} = \frac{(1-D/P)KC[(1-N/100)(K_1+I)-(1-M/100)(K+I)]}{KC[(1-M/100)(K+I)-(1-N/100)(1-D/P)\{K(1-M/100)+I\}]}$$

Or $$R = \frac{100(1-D/P)[(1-N/100)\{K(1-M/100)+I\}-(1-M/100)(K+I)]}{[(1-M/100)(K+I)-(1-N/100)(1-D/P)\{K(1-M/100)+I\}]}$$

With the reference parameters:

D	*C*	*I*	*P*	*K*	*T*	*E*
600	50	35	960	100	0.131	763.76

For example, if $M = 15$ then:

$$R = \frac{4.375-N}{1.55+0.01N}$$

For $M = 15$, the various combinations of R and N are given as follows:

S. No.	*N*	$R = \frac{4.375-N}{1.55+0.01N}$
1	1	2.16
2	2	1.51
3	3	0.87
4	4	0.24

7.5.2 Shorter cycle time

Cycle time decreases also because of an increase in shortage cost. For a similar cycle time, consider the management response in the form of:

(i) Reduction in the holding cost

(ii) Reduction in the production rate

It can be analysed independently. However as discussed before, a combination of parameters may be also useful.

Combination of holding cost and production rate:

Let M, N, and R be the % variation in shortage cost, holding cost, and production rate.

Now,

$$K_1 = K\left(1+\frac{M}{100}\right)$$

$$I_1 = I\left(1-\frac{N}{100}\right)$$

$$P_1 = P\left(1-\frac{R}{100}\right)$$

For a similar cycle time:

$$\sqrt{\frac{2C(K+I)}{KID(1-D/P)}} = \sqrt{\frac{2C(K_1+I_1)}{K_1I_1D(1-D/P_1)}}$$

Or $$\frac{(K+I)}{KI(1-D/P)} = \frac{(K_1+I_1)}{K_1I_1(1-D/P_1)}$$

Or $$1-\frac{D}{P_1} = \frac{KI(K_1+I_1)(1-D/P)}{K_1\ I_1\ (K+I)}$$

Or $$\frac{D}{P_1} = \frac{K_1I_1(K+I)-KI(K_1+I_1)(1-D/P)}{K_1\ I_1\ (K+I)}$$

Or $$\frac{P_1}{D} = \frac{K_1I_1(K+I\ )}{K_1I_1(K+I)-KI(K_1+I_1)(1-D/P)}$$

Or $$1-\frac{R}{100} = \frac{(D/P)K_1I_1(K+I\ )}{K_1I_1(K+I)-KI(K_1+I_1)(1-D/P)}$$

Or $$\frac{R}{100} = \frac{K_1I_1(K+I)-KI(K_1+I_1)(1-D/P)-(D/P)K_1I_1(K+I)}{K_1I_1(K+I)-KI(K_1+I_1)(1-D/P)}$$

Or $$\frac{R}{100} = \frac{K_1I_1(K+I)(1-D/P)-KI(K_1+I_1)(1-D/P)}{K_1I_1\ (K+I)-KI(K_1+I_1)(1-D/P)}$$

Or $$\frac{R}{100} = \frac{(1-D/P)[K_1I_1(K+I)-KI(K_1+I_1)]}{K_1I_1(K+I)-KI(K_1+I_1)(1-D/P)}$$

Or $$\frac{R}{100} = \frac{(1-D/P)[(1+M/100)(1-N/100)(K+I)-(K_1+I_1)]}{(1+M/100)(1-N/100)(K+I)-(K_1+I_1)(1-D/P)}$$

Or $$R = \frac{100(1-D/P)[(1+M/100)(1-N/100)(K+I)-\{K(1+M/100) + I(1-N/100)\}]}{(1+M/100)(1-N/100)(K+I)-[(1-D/P)\{K(1+M/100) + I(1-N/100)\}]}$$

With the reference parameters:

D	*C*	*I*	*P*	*K*	*T*	*E*
600	50	35	960	100	0.131	763.76

For example, if $M = 15$, then

$$R = \frac{65.625 - 15.03125N}{33 - 0.47375N}$$

For $M = 15$, the various combinations of N and R are given as follows:

S. No.	*N*	$R = \frac{65.625 - 15.03125N}{33 - 0.47375N}$
1	1	1.55
2	2	1.11
3	3	0.65
4	4	0.18

8
Shortages

Abstract : There are instances when shortages are being faced by the production organisation and the cost is associated to this. In order to estimate such cost, it is necessary to understand the reasons for occurrence of shortages. This chapter explains various such reasons in addition to the effects on the output parameters. The relevant output parameters are batch size, shortage quantity and total cost. Effects on these output parameters are reported with respect to the stock out or shortage cost increase and decrease. When a change is initiated by the shortage cost, the possible management response is also studied.

Keyword : Shortages, reasons for shortages, shortage cost, output parameters, management response

There are situations when demand of a product is available, although existing stock is zero. In such a case also, however, if production system is not operational concerning similar product for certain period, then the shortages are said to occur.

8.1 Shortage cost estimation

In order to estimate such cost, it is necessary to understand the reasons for occurrence of shortages. Shortages may occur because of various reasons in different context pertaining to production environment. Some of these reasons (Fig. 8.1) are as follows:

(a) Raw material or component is not available for certain period and therefore the production facility is idle.

(b) A worker on machine is not available for some period and therefore the machine is not functional

(c) Because of maintenance problems or breakdown, a machine cannot work.

(d) Owing to unavoidable space constraint on the shop floor for certain period, production is stopped. This might be due to a difficulty in keeping the work-in-process (WIP) temporarily.

(e) Priority is given to some other job/task on a facility and therefore the product/component under consideration could not be processed for certain period.

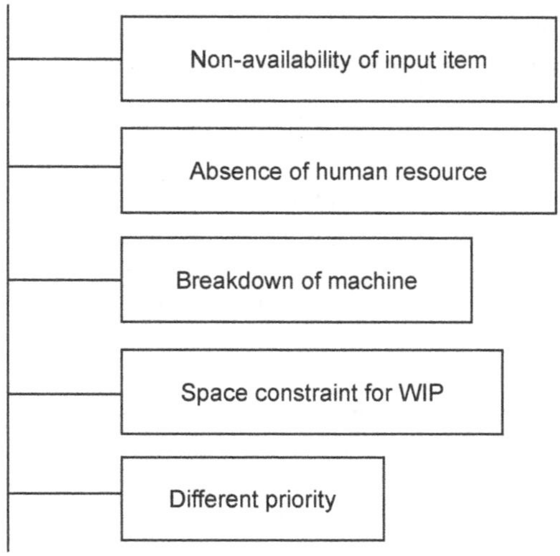

Figure 8.1: Factors associated with shortages

In order to estimate the shortage cost, the effects of shortages should be quantified in terms of cost. Some of the potential effects are as given below:

(i) Because of maintenance problems or breakdown, the relevant worker is not engaged in productive activity leading to a loss in terms of wages/salary for certain period.

(ii) Component at an intermediate stage is facing shortages. All the subsequent processes could not be performed for some period leading to considerable cost. In such a case, effects have gone beyond a facility under consideration.

(iii) Since finished product could not reach the customer on time, there might be certain dissatisfaction. Although it is backordered this time, however, it might affect future potential orders of a customer company or demand of individual customer.

Shortage or stock out cost needs to be estimated per unit product for a specified period such as an annual shortage cost per unit item. When all the shortages are allowed to be backordered, the output parameters have been provided before.

Example 8.1

Consider the following parameters:

Annual demand, $D = 600$ units

Setup cost, $C =$ ₹ 45

Annual production rate, $P = 960$ units

Annual inventory carrying cost per unit, $I =$ ₹ 40

Annual shortage cost per unit, $K =$ ₹ 100

Now:

Optimum production batch quantity:

$$Q^* = \sqrt{\frac{2DC(K+I)}{KI(1-D/P)}}$$

$$= 71 \text{ Units}$$

An optimum shortage quantity:

$$J^* = \sqrt{\frac{2DCI(1-D/P)}{K(K+I)}}$$

$$= 7.61 \text{ units}$$

And the total annual cost:

$$E^* = \sqrt{\frac{2DCKI(1-D/P)}{(K+I)}}$$

$$= ₹\ 760.64$$

It is of interest to examine the shortage cost variation.

8.2 Stock out cost increase

Shortage or stock out cost increases because of the reasons such as:

(i) If experienced employees are idle for want of input items or breakdown of machine, then the corresponding shortage cost might increase because of higher wages/salaries.

(ii) In case where a particular facility is used for certain highly profitable product and because of its stoppage, stock out occurs, then the associated cost increases due to higher profit loss.

(iii) When extended effects of shortages are observed beyond a particular facility and for longer period, then the shortage or stock out cost estimation is on higher side.

Example 8.2

With the use of the following information from the previous example:

P	*D*	*C*	*I*	*K*	*Q*	*E*	*J*
960	600	45	40	100	71.0	760.64	7.61

Effects of shortage cost are examined considering its variation as:

% Increase in K	5%	10%	15%	20%	25%	30%
K	105	110	115	120	125	130

Table 8.1 shows the corresponding variation in output parameters. Batch size and shortage quantity decrease whereas total cost increases.

Table 8.1: Effects on output parameters related to shortage cost increase.

% Increase in K	5%	10%	15%	20%	25%	30%
K	105	110	115	120	125	130
Q	70.508	70.065	69.658	69.282	68.935	68.613
% Decrease in Q	0.68%	1.31%	1.88%	2.41%	2.90%	3.35%
E	765.87	770.71	775.22	779.42	783.35%	787.03
% Increase in E	0.69%	1.32%	1.92%	2.47%	2.99%	3.47%
J	7.29	7.01	6.74	6.50	6.27	6.05
% Decrease in J	4.11%	7.89%	11.38%	14.61%	17.61%	20.41%

In order to generalise: $K_1 = K\left(1+\frac{M}{100}\right)$

(i) Reduction in production batch size

$$= \sqrt{\frac{2DC(K+I)}{KI(1-D/P)}} - \sqrt{\frac{2DC(K_1+I)}{K_1 I(1-D/P)}}$$

$$= \sqrt{\frac{2DC(K+I)}{KI(1-D/P)}}\left[1-\sqrt{\frac{K(K_1+I)}{K_1(K+I)}}\right]$$

$$= \sqrt{\frac{2DC(K+I)}{KI(1-D/P)}}\left[1-\sqrt{\frac{K(1+M/100)+I}{(1+M/100)(K+I)}}\right]$$

And:

$$\%\text{reduction in } Q = 1-\sqrt{\frac{K(1+M/100)+I}{(1+M/100)(K+I)}}$$

(ii) Increase in the total cost

$$= \sqrt{\frac{2DCK_1 I(1-D/P)}{(K_1+I\ )}} - \sqrt{\frac{2DCKI(1-D/P)}{(K+I)}}$$

$$= \sqrt{\frac{2DCKI(1-D/P)}{(K+I\ )}}\left[\sqrt{\frac{K_1(K+I)}{K(K_1+I)}}-1\right]$$

$$= \sqrt{\frac{2DCKI(1-D/P)}{(K+I\ )}}\left[\sqrt{\frac{(1+M/100)(K+I)}{K(1+M/100)+I}}-1\right]$$

And:

$$\%\text{ increase in } E = \sqrt{\frac{(1+M/100)(K+I)}{K(1+M/100)+I}}-1$$

(iii) Decrease in shortage quantity

$$= \sqrt{\frac{2DCI(1-D/P)}{K(K+I\ )}} - \sqrt{\frac{2DCI(1-D/P)}{K_1(K_1+I)}}$$

$$= \sqrt{\frac{2DCI(1-D/P)}{K(K+I\ )}}\left[1-\sqrt{\frac{K(K+I)}{K_1(K_1+I)}}\right]$$

$$= \sqrt{\frac{2DCI(1-D/P)}{K(K+I\ )}}\left[1-\sqrt{\frac{(K+I)}{(1+M/100)\{K(1+M/100)+I\}}}\right]$$

And:

$$\%\text{ decrease in } J = 1-\sqrt{\frac{(K+I)}{(1+M/100)\{K(1+M/100)+I\}}}$$

Table 8.2 shows the summarised results.

Table 8.2: Results with respect to % stock out cost increase.

Reduction in the production batch size	$\sqrt{\frac{2DC(K+I)}{KI(1-D/P)}}\left[1-\sqrt{\frac{K(1+M/100)+I}{(1+M/100)(K+I)}}\right]$
% reduction in the production batch size	$1-\sqrt{\frac{K(1+M/100)+I}{(1+M/100)(K+I)}}$
Increase in the total cost	$\sqrt{\frac{2DCKI(1-D/P)}{(K+I)}}\left[\sqrt{\frac{(1+M/100)(K+I)}{K(1+M/100)+I}}-1\right]$
% increase in the total cost	$\sqrt{\frac{(1+M/100)(K+I)}{K(1+M/100)+I}}-1$
Decrease in the shortage quantity	$\sqrt{\frac{2DCI(1-D/P)}{K(K+I\ )}}\left[1-\sqrt{\frac{(K+I)}{(1+M/100)\{K(1+M/100)+I\}}}\right]$
%decrease in J	$1-\sqrt{\frac{(K+I)}{(1+M/100)\{K(1+M/100)+I\}}}$

Refer Table 8.1. Percentage variation in production batch size is lower than that in the shortage quantity. This is because

$$1-\sqrt{\frac{K(1+M/100)+I}{(1+M/100)(K+I)}}<1-\sqrt{\frac{(K+I)}{(1+M/100)\{K(1+M/100)+I\}}}$$

Or $$\sqrt{\frac{K(1+M/100)+I}{(1+M/100)(K+I)}}>\sqrt{\frac{(K+I)}{(1+M/100)\{K(1+M/100)+I\}}}$$

Or $$\sqrt{\frac{K(1+M/100)+I}{(K+I)}}>\sqrt{\frac{(K+I)}{\{K(1+M/100)+I\}}}$$

Or $K(1+M/100)+I>(K+I)$

Or $K(1+M/100)>K$

And this is true for all practical values of M, i.e., $0>M<100$.

8.3 Stock out cost decrease

Shortage cost reduces because of the reasons such as:

(a) In case of a new company or otherwise, if less experienced and young workers are associated with a particular machine, their idle time cost would be lower due to lower wages/salaries.

(b) If profit margin on certain product is lower, shortage cost would be lower in case of stock out because profit loss is lower per unit item.

(c) When effects of shortages are not extended or an effort is made to localize the effects, the shortage cost estimation is expected to be on lower side.

Example 8.3

With the use of the following information:

P	*D*	*C*	*I*	*K*	*Q*	*E*	*J*
960	600	45	40	100	71.0	760.64	7.61

Effects of shortage cost are examined considering its variation as:

% Decrease in K	5%	10%	15%	20%	25%	30%
K	95	90	85	80	75	70

Table 8.3 shows the corresponding variation in output parameters. Batch size and shortage quantity increase whereas total cost reduces.

Table 8.3: Effects on output parameters related to shortage cost decrease.

% Decrease in K	5%	10%	15%	20%	25%	30%
K	95	90	85	80	75	70
Q	71.525	72.111	72.761	73.485	74.297	75.214
% Increase in Q	0.75%	1.57%	2.49%	3.51%	4.65%	5.95%
E	754.98	748.85	742.16	734.85	726.82	717.95
% Decrease in E	0.74%	1.55%	2.43%	3.39%	4.45%	5.61%
J	7.95	8.32	8.73	9.19	9.69	10.26
% Increase in J	4.48%	9.39%	14.79%	20.76%	27.40%	34.84%

In order to generalize:

$$K_1 = K\left(1-\frac{M}{100}\right)$$

(i) Increase in production batch size

$$= \sqrt{\frac{2DC(K_1+I)}{K_1I(1-D/P)}} - \sqrt{\frac{2DC(K+I)}{KI(1-D/P)}}$$

$$= \sqrt{\frac{2DC(K+I)}{KI(1-D/P)}}\left[\sqrt{\frac{K(K_1+I)}{K_1(K+I)}}-1\right]$$

$$= \sqrt{\frac{2DC(K+I)}{KI(1-D/P)}}\left[\sqrt{\frac{K(1-M/100)+I}{(1-M/100)(K+I)}}-1\right]$$

And:

$$\%\text{ increase in } Q = \sqrt{\frac{K(1-M/100)+I}{(1-M/100)(K+I)}}-1$$

(ii) Reduction in the total cost

$$= \sqrt{\frac{2DCKI(1-D/P)}{(K+I)}} - \sqrt{\frac{2DCK_1I(1-D/P)}{(K_1+I)}}$$

$$= \sqrt{\frac{2DCKI(1-D/P)}{(K+I)}}\left[1-\sqrt{\frac{K_1(K+I)}{K(K_1+I)}}\right]$$

$$= \sqrt{\frac{2DCKI(1-D/P)}{(K+I)}}\left[1-\sqrt{\frac{(1-M/100)(K+I)}{K(1-M/100)+I}}\right]$$

And:

$$\%\text{reduction } E = 1-\sqrt{\frac{(1-M/100)(K+I)}{K(1-M/100)+I}}$$

(iii) Increase in shortage quantity

$$= \sqrt{\frac{2DCI(1-D/P)}{K_1(K_1+I)}} - \sqrt{\frac{2DCI(1-D/P)}{K(K+I)}}$$

$$= \sqrt{\frac{2DCI(1-D/P)}{K(K+I)}}\left[\sqrt{\frac{K(K+I)}{K_1(K_1+I)}}-1\right]$$

$$= \sqrt{\frac{2DCI(1-D/P)}{K(K+I)}}\left[\sqrt{\frac{(K+I)}{(1-M/100)\{K(1-M/100)+I\}}}-1\right]$$

And:

$$\% \text{ increase in } J = \sqrt{\frac{(K+I)}{(1-M/100)\{K(1-M/100)+I\}}}-1$$

Table 8.4 shows the summarized results.

Table 8.4: Result with respect to % stock out cost decrease.

Increase in the production batch size	$\sqrt{\frac{2DC(K+I)}{KI(1-D/P)}}\left[\sqrt{\frac{K(1-M/100)+I}{(1-M/100)(K+I)}}-1\right]$
% Increase in the production batch size	$\sqrt{\frac{K(1-M/100)+I}{(1-M/100)(K+I)}}-1$
Reduction in the total cost	$\sqrt{\frac{2DCKI(1-D/P)}{(K+I)}}\left[1-\sqrt{\frac{(1-M/100)(K+I)}{K(1-M/100)+I}}\right]$
% Reduction in the total cost	$1-\sqrt{\frac{(1-M/100)(K+I)}{K(1-M/100)+I}}$
Increase in the shortage Quantity	$\sqrt{\frac{2DCI(1-D/P)}{K(K+I)}}\left[\sqrt{\frac{(K+I)}{(1-M/100)\{K(1-M/100)+I\}}}-1\right]$
% Increase in J	$\sqrt{\frac{(K+I)}{(1-M/100)\{K(1-M/100)+I\}}}-1$

Refer Table 8.3. Percentage variation in shortage quantity is greater than that in the production batch size. This is because:

$$\sqrt{\frac{(K+I)}{(1-M/100)\{K(1-M/100)+I\}}}-1 > \sqrt{\frac{K(1-M/100)+I}{(1-M/100)(K+I)}}-1$$

Or $$\sqrt{\frac{(K+I)}{(1-M/100)\{K(1-M/100)+I\}}} > \sqrt{\frac{K(1-M/100)+I}{(1-M/100)(K+I)}}$$

Or $$\sqrt{\frac{(K+I)}{\{K(1-M/100)+I\}}} > \sqrt{\frac{K(1-M/100)+I}{(K+I)}}$$

Or $(K+I) > K(1-M/100)+I$

Or $K > K(1-M/100)$

And this is true for all practical values of M, i.e., $0 > M < 100$.

8.4 Interaction with the parameters

When change is initiated by the shortage cost, the possible management response is studied.

8.4.1 Shortage cost reduction

In case of shortage cost reduction, setup cost reduction and also production rate increase might be explored for an objective of similar batch size.

8.4.1.1 Setup cost reduction

For a general approach, let:

M = % decrease in shortage cost

N = % reduction in setup cost

Therefore:

$$K_1 = K\left(1-\frac{M}{100}\right)$$

$$C_1 = C\left(1-\frac{N}{100}\right)$$

For similar batch size:

$$\sqrt{\frac{2DC(K+I)}{KI(1-D/P)}} = \sqrt{\frac{2DC(K_1+I)}{K_1I(1-D/P)}}$$

Or $$\frac{C(K+I)}{K} = \frac{C_1(K_1+I)}{K_1}$$

Or $$\frac{C(K+I)}{K} = C_1\left(1+\frac{I}{K_1}\right)$$

Or $$\frac{(K+I)}{K(1-N/100)} = 1+\frac{I}{K(1-M/100)}$$

$$\text{Or} \quad \frac{(K+I)}{K(1-N/100)} = \frac{K(1-M/100)\ +I}{K(1-M/100)}$$

$$\text{Or} \quad \frac{(1-N/100)}{(K+I)} = \frac{(1-M/100)}{K(1-M/100)+I}$$

$$\text{Or} \quad 1-\frac{N}{100} = \frac{(1-M/100)(K+I)}{K(1-M/100)+I}$$

$$\text{Or} \quad \frac{N}{100} = \frac{K(1-M/100)+I-(1-M/100)K-(1-M/100)I}{K(1-M/100)+I}$$

$$\text{Or} \quad \frac{N}{100} = \frac{I(1-1+M/100)}{K(1-M/100)+I}$$

$$\text{Or} \quad \frac{N}{100} = \frac{I(M/100)}{K(1-M/100)+I}$$

$$\text{Or} \quad N = \frac{IM}{K(1-M/100)+I}$$

With the reference set of data as follows:

P	D	C	I	K	Q	E	J
960	600	45	40	100	71.0	760.64	7.61

Values of N are represented by Table 8.5. These values are lower than that of M but are more sensitive toward higher values of M.

Table 8.5: Variation of N (setup cost) corresponding to M (reduced K).

Sr. no	M	$N = \frac{IM}{K(1-M/100)+I}$
1	10	3.08
2	15	4.8
3	20	6.67
4	25	8.70
5	30	10.91

8.4.1.2 Production rate increase

For a general approach, let:

M = % decrease in shortage cost

N = % increase in production rate

Therefore,

$$K_1 = K\left(1-\frac{M}{100}\right)$$

$$P_1 = P\left(1+\frac{N}{100}\right)$$

For similar batch size:

$$\sqrt{\frac{2DC(K+I)}{KI(1-D/P)}} = \sqrt{\frac{2DC(K_1+I)}{K_1 I(1-D/P_1)}}$$

Or $$\frac{(K+I)}{K(1-D/P)} = \frac{(K_1+I)}{K_1(1-D/P_1)}$$

Or $$\frac{(K+I)}{(1-D/P)} = \frac{(K_1+I)}{(1-M/100)(1-D/P_1)}$$

Or $$1-\frac{D}{P_1} = \frac{[K(1-M/100)+I](1-D/P)}{(1-M/100)\ (K+I)}$$

Or $$\frac{D}{P_1} = \frac{(1-M/100)(K+I)-K(1-M/100)(1-D/P)-I(1-D/P)}{(1-M/100)(K+I)}$$

Or $$P_1 = \frac{D(1-M/100)(K+I)}{(1-M/100)(K+I)-K(1-M/100)(1-D/P)-I(1-D/P)}$$

Or $$1+\frac{N}{100} = \frac{(D/P)(1-M/100)(K+I)}{(1-M/100)I+K(1-M/100)(D/P)-I\ (1-D/P)}$$

Or $$\frac{N}{100} = \frac{(D/P)(1-M/100)(K+I)-(1-M/100)I-K(1-M/100)(D/P)+I(1-D/P)}{(1-M/100)I+K(1-M/100)(D/P)-I(1-D/P)}$$

Or $$\frac{N}{100} = \frac{(D/P)(1-M/100)I-(1-M/100)I+I(1-D/P)}{K(1-M/100)(D/P)-(IM/100)+(ID/P)}$$

Or $$\frac{N}{100} = \frac{(ID/P)-(ID/P)(M/100)-I+(IM/100)+I-(ID/P)}{K(1-M/100)(D/P)-(IM/100)+(ID/P)}$$

Or $$\frac{N}{100} = \frac{(IM/100)(1-D/P)}{(K+I)(D/P)-(M/100)\{I+K(D/P)\}}$$

Or $$N = \frac{IM(1-D/P)}{(K+I)(D/P)-(M/100)\{I+K(D/P)\}}$$

With the reference set of data as follows:

P	D	C	I	K	Q	E	J
960	600	45	40	100	71.0	760.64	7.61

Values of N are shown in Table 8.6. These values are lower than that of M but are more sensitive towards higher values of M.

Table 8.6: Variation of N (production rate) corresponding to M (reduced K).

S. No.	M	$N = \frac{IM(1-D/P)}{(K+I)(D/P)-(M/100)\{I+K(D/P)\}}$
1	10	1.94
2	15	3.12
3	20	4.48
4	25	6.06
5	30	7.93

8.4.2 Shortage cost increase

In case of the shortage cost increase, the following objectives may be explored:

(i) Similar production lot size

(ii) Similar total annual cost

8.4.2.1 Similar production lot size

Management response can be in the form of demand increase/holding cost reduction/production rate decrease.

(a) Demand increase

$$K_1 = K\left(1+\frac{M}{100}\right)$$

$$D_1 = D\left(1+\frac{N}{100}\right)$$

$$\sqrt{\frac{2DC(K+I)}{KI(1-D/P)}} = \sqrt{\frac{2D_1C(K_1+I)}{K_1I(1-D_1/P)}}$$

Or $$\frac{D(K+I)}{K(1-D/P)} = \frac{D_1(K_1+I)}{K_1(1-D_1/P)}$$

$$\text{Or} \quad \frac{(K+I)}{(1-D/P)} = \frac{(1+N/100)(K_1+I)}{(1+M/100)(1-D_1/P)}$$

$$\text{Or} \quad \frac{(K+I)}{(1-D/P)} = \frac{(1+N/100)\{K(1+M/100)+I\}}{(1+M/100)\{1-(D/P)(1+N/100)\}}$$

$$\text{Or} \quad \frac{1-(D/P)(1+N/100)}{(1+N/100)} = \frac{(1-D/P)\{K(1+M/100)+I\}}{(K+I)(1+M/100)}$$

$$\text{Or} \quad \frac{1}{(1+N/100)} = \frac{(D/P)(K+I)(1+M/100)+(1-D/P)\{K(1+M/100)+I\}}{(K+I)(1+M/100)}$$

$$\text{Or} \quad \frac{1}{(1+N/100)} = \frac{(D/P)I(1+M/100)+K(1+M/100)+I-(ID/P)}{(K+I)(1+M/100)}$$

$$\text{Or} \quad \frac{1}{(1+N/100)} = \frac{\left(\frac{D}{P}\right)I\left(\frac{M}{100}\right)+(K+I)+(KM/100)}{(K+I)(1+M/100)}$$

$$\text{Or} \quad 1+\frac{N}{100} = \frac{(K+I)(1+M/100)}{(D/P)(IM/100)+(K+I)+(KM/100)}$$

$$\text{Or} \quad \frac{N}{100} = \frac{(K+I)(1+M/100)-(D/P)(IM/100)-K-I-(KM/100)}{(M/100)\{K+(ID/P)\}+(K+I)}$$

$$\text{Or} \quad \frac{N}{100} = \frac{K+(KM/100)+I+(IM/100)-(D/P)(IM/100)-K-I-(KM/100)}{(M/100)\{K+(ID/P)\}+(K+I)}$$

$$\text{Or} \quad \frac{N}{100} = \frac{(IM/100)(1-D/P)}{(M/100)\{K+(ID/P)\}+(K+I)}$$

$$\text{Or} \quad N = \frac{IM(1-D/P)}{(M/100)\{K+(ID/P)\}+(K+I)}$$

With the reference set of data as follows:

P	*D*	*C*	*I*	*K*	*Q*	*E*	*J*
960	600	45	40	100	71.0	760.64	7.61

Values of *N* are represented by Table 8.7. These values are much lower than that of *M*, and are also less sensitive towards higher values of *M*.

Table 8.7: Variation of *N* (demand) corresponding to *M* (increased *K*).

S. No.	M	$N = \dfrac{IM(1-D/P)}{\left(\dfrac{M}{100}\right)\left\{K+\left(\dfrac{ID}{P}\right)\right\}+(K+I)}$
1	10	0.98
2	15	1.42
3	20	1.82
4	25	2.19
5	30	2.54

(b) Inventory holding cost reduction

$$K_1 = K\left(1+\frac{M}{100}\right)$$

$$I_1 = I\left(1-\frac{N}{100}\right)$$

Now:

$$\sqrt{\frac{2DC(K+I)}{KI(1-D/P)}} = \sqrt{\frac{2DC(K_1+I_1)}{K_1I_1\ (1-D/P)}}$$

Or $$\frac{(K+I)}{KI} = \frac{(K_1+I_1)}{K_1I_1} \qquad (8.1)$$

Or $$\frac{(K+I)}{KI} = \frac{1}{I_1}+\frac{1}{K_1}$$

Or $$\frac{1}{I_1} = \frac{(K+I)}{KI}-\frac{1}{K_1}$$

Or $$\frac{1}{I_1} = \frac{K_1(K+I)-KI}{KK_1I}$$

Or $$1-\frac{N}{100} = \frac{KK_1}{K_1(K+I)-KI}$$

Or $$\frac{N}{100} = \frac{K_1(K+I)-KI-KK_1}{K_1(K+I)-KI}$$

Or $$\frac{N}{100} = \frac{I(1+M/100)-I}{K(1+M/100)+(IM/100)}$$

Or $$\frac{N}{100} = \frac{(IM/100)}{K+(K+I)(M/100)}$$

Or $$N = \frac{IM}{K + (K + I)(M/100)}$$

With the reference set of data as before, values of N are shown in Table 8.8 considering relevant parameters:

$$K = 100$$
$$I = 40$$

Values of N are lower than that of M and also less sensitive towards higher values of M.

Table 8.8: Variation of N (holding cost) corresponding to M (increased K).

S. No.	M	$N = \frac{IM}{K + (K + I)(M/100)}$
1	10	3.51
2	15	4.96
3	20	6.25
4	25	7.41
5	30	8.45

(c) Production rate decrease

$$K_1 = K\left(1 + \frac{M}{100}\right)$$

$$P_1 = P\left(1 - \frac{N}{100}\right)$$

Now:

$$\sqrt{\frac{2DC(K + I)}{KI(1 - D/P)}} = \sqrt{\frac{2DC(K_1 + I)}{K_1 I(1 - D/P_1)}}$$

Or $$\frac{(K + I)}{K(1 - D/P)} = \frac{(K_1 + I)}{K_1(1 - D/P_1)} \quad (8.2)$$

Or $$1 - \frac{D}{P_1} = \frac{K(1 - D/P)(K_1 + I)}{K_1(K + I)}$$

Or $$1 - \frac{D}{P_1} = \frac{(1 - D/P)(K_1 + I)}{(1 + M/100)(K + I)}$$

Or $$\frac{D}{P_1} = \frac{(1 + M/100)(K + I) - (1 - D/P)\ (K_1 + I)}{(1 + M/100)(K + I)}$$

$$\text{Or} \quad P_1 = \frac{D(1+M/100)(K+I)}{(1+M/100)(K+I)-(1-D/P)\{K(1+M/100)+I\}}$$

$$\text{Or} \quad 1-\frac{N}{100} = \frac{(D/P)(1+M/100)(K+I)}{(1+M/100)I-I+(D/P)K(1+M/100)+(D/P)I}$$

$$\text{Or} \quad 1-\frac{N}{100} = \frac{(D/P)(1+M/100)(K+I)}{(IM/100)+(D/P)(K+I)+(D/P)(KM/100)}$$

$$\text{Or} \quad \frac{N}{100} = \frac{(IM/100)+(D/P)(K+I)+(D/P)(KM/100)-(D/P)(1+M/100)(K+I)}{(M/100)(I+KD/P)+(D/P)(K+I)}$$

$$\text{Or} \quad \frac{N}{100} = \frac{(IM/100)+(D/P)(KM/100)-(D/P)(M/100)(K+I)}{(M/100)(I+KD/P)+(D/P)(K+I)}$$

$$\text{Or} \quad N = \frac{IM+(D/P)(KM)-(D/P)(M)(K+I)}{(M/100)(I+KD/P)+(D/P)(K+I)}$$

$$\text{Or} \quad N = \frac{IM-(D/P)IM}{(M/100)(I+KD/P)+(D/P)(K+I)}$$

$$\text{Or} \quad N = \frac{IM(1-D/P)}{(M/100)(I+KD/P)+(D/P)(K+I)}$$

With the reference set of data as follows

P	D	C	I	K	Q	E	J
960	600	45	40	100	71.0	760.64	7.61

Values of N are shown in Table 8.9. These values are lower than that of M and are also less sensitive towards higher values of M.

Table 8.9: Variation of N (production rate) corresponding to M (increased K).

S. No.	M	$N = \frac{IM(1-D/P)}{(M/100)(I+KD/P)+(D/P)(K+I)}$
1	10	1.53
2	15	2.19
3	20	2.78
4	25	3.31
5	30	3.81

8.4.2.2 Similar total annual cost

With the objective of similar total annual cost, a management response can be in the form of setup cost reduction/production rate reduction/holding cost decrease.

(a) Setup cost reduction

$$K_1 = K\left(1+\frac{M}{100}\right)$$

$$C_1 = C\left(1-\frac{N}{100}\right)$$

Now:

$$\sqrt{\frac{2DCKI(1-D/P)}{(K+I)}} = \sqrt{\frac{2DC_1K_1I(1-D/P)}{(K_1+I)}}$$

Or $$\frac{CK}{(K+I)} = \frac{C_1K_1}{(K_1+I)}$$

Or $$C_1 = \frac{CK(K_1+I)}{K_1(K+I)}$$

Or $$1-\frac{N}{100} = \frac{K(1+M/100)+I}{(1+M/100)(K+I)}$$

Or $$\frac{N}{100} = \frac{(1+M/100)(K+I)-K(1+M/100)-I}{(1+M/100)(K+I)}$$

Or $$\frac{N}{100} = \frac{I(1+M/100)-I}{(1+M/100)(K+I)}$$

Or $$N = \frac{IM}{(1+M/100)(K+I)}$$

With the reference set of data as before, values of N are shown in Table 8.10 considering relevant parameters:

$$K = 100$$

$$I = 40$$

Values of N are lower than that of M and also slightly less sensitive towards higher values of M.

Table 8.10: Variation of *N* (setup cost) corresponding to *M* (increased *K*)

S. No.	*M*	$N = \dfrac{IM}{(1+M/100)(K+I)}$
1	10	2.60
2	15	3.73
3	20	4.76
4	25	5.71
5	30	6.59

(b) Production rate reduction

$$K_1 = K\left(1+\frac{M}{100}\right)$$

$$P_1 = P\left(1-\frac{N}{100}\right)$$

Now:

$$\sqrt{\frac{2DCKI(1-D/P)}{(K+I)}} = \sqrt{\frac{2DCK_1I(1-D/P_1)}{(K_1+I)}}$$

Or

$$\frac{K(1-D/P)}{(K+I)} = \frac{K_1(1-D/P_1)}{(K_1+I)}$$

As it is similar to Expression (8.2), the result can similarly be obtained as:

$$N = \frac{IM(1-D/P)}{(M/100)(I+KD/P)+(D/P)(K+I)}$$

(c) Holding cost decrease

$$K_1 = K\left(1+\frac{M}{100}\right)$$

$$I_1 = I\left(1-\frac{N}{100}\right)$$

Now:

$$\sqrt{\frac{2DCKI(1-D/P)}{(K+I)}} = \sqrt{\frac{2DCK_1I_1(1-D/P)}{(K_1+I_1)}}$$

Or

$$\frac{KI}{(K+I)} = \frac{K_1I_1}{(K_1+I_1)}$$

As it similar to Expression (8.1), the result can similarly be obtained as:

$$N = \frac{IM}{K+(K+I)(M/100)}$$

8.5 Variation of K triggered by change in other parameters

Change in parameters other than the shortage cost (K) is imposed and response as the variation in K is examined.

In case of demand decrease, shortage cost reduction can be a potential response for similar batch size.

$$D_1 = D\left(1-\frac{M}{100}\right)$$

$$K_1 = K\left(1-\frac{N}{100}\right)$$

And:

$$\sqrt{\frac{2DC(K+I)}{KI(1-D/P)}} = \sqrt{\frac{2D_1C(K_1+I)}{K_1I(1-D_1/P)}}$$

Or $$\frac{D(K+I)}{K(1-D/P)} = \frac{D_1(K_1+I)}{K_1\ (1-D_1/P)}$$

Or $$\frac{K_1+I}{K_1} = \frac{D(K+I)(1-D_1/P)}{D_1K(1-D/P)}$$

Or $$\frac{K_1+I}{K_1} = \frac{(K+I)(1-D_1/P)}{K(1-M/100)(1-D/P)}$$

Or $$\frac{I}{K_1} = \frac{(K+I)\{1-(D/P)(1-M/100)\}-K(1-M/100)(1-D/P)}{K(1-M/100)(1-D/P)}$$

Or $$\frac{I}{(1-N/100)} = \frac{(K+I)-(K+I)(D/P)(1-M/100) - K(1-M/100)(1-D/P)}{(1-M/100)(1-D/P)}$$

Or $$1-\frac{N}{100} = \frac{I(1-M/100)(1-D/P)}{(K+I)-K(D/P)(1-M/100)-I(D/P)(1-M/100) - K(1-M/100)+K(1-M/100)(D/P)}$$

Or $$1-\frac{N}{100} = \frac{I(1-M/100)(1-D/P)}{(K+I)-I(D/P)(1-M/100)-K(1-M/100)}$$

Or $$\frac{N}{100} = \frac{(K+I)-I(D/P)(1-M/100)-K(1-M/100) - I(1-M/100)(1-D/P)}{I-I(D/P)(1-M/100)+(KM/100)}$$

$$\text{Or} \quad \frac{N}{100} = \frac{(K+I)-K(1-M/100)-I(1-M/100)}{I-I(D/P)+I(D/P)(M/100)+(KM/100)}$$

$$\text{Or} \quad \frac{N}{100} = \frac{I+(KM/100)-I+(IM/100)}{I(1-D/P)+(M/100)\{K+(ID/P)\}}$$

$$\text{Or} \quad \frac{N}{100} = \frac{(K+1)(M/100)}{I(1-D/P)+(M/100)\{K+(ID/P)\}}$$

$$\text{Or} \quad N = \frac{M(K+1)}{I(1-D/P)+(M/100)\{K+(ID/P)\}}$$

With the reference set of data as follows:

P	*D*	*C*	*I*	*K*	*Q*	*E*	*J*
960	600	45	40	100	71.0	760.64	7.61

Values of N are shown in Table 8.11. These are much higher than that of M but are less sensitive towards higher values of M.

Table 8.11: Variation of N (shortage cost) corresponding to M (decreased D).

S. No.	*M*	$N = \frac{M(K+I)}{I(1-D/P)+(M/100)\{K+(ID/P)\}}$
1	2	16.00
2	4	28.00
3	6	37.33
4	8	44.80
5	10	50.91

8.5.2 Setup cost

In case of setup cost decrease, a possible response for similar batch size can be the shortage cost reduction.

$$C_1 = C\left(1-\frac{M}{100}\right)$$

$$K_1 = K\left(1-\frac{N}{100}\right)$$

And:

$$\sqrt{\frac{2DC(K+I)}{KI(1-D/P)}} = \sqrt{\frac{2DC_1(K_1+I)}{K_1 I(1-D/P)}}$$

Or $$\frac{C(K+I)}{K}=\frac{C_1(K_1+I)}{K_1}$$

Or $$\frac{(K_1+I)}{K_1}=\frac{C(K+I)}{C_1K}$$

Or $$\frac{(K_1+I)}{K_1}=\frac{(K+I)}{K(1-M/100)}$$

Or $$\frac{I}{K_1}=\frac{(K+I)-K(1-M/100)}{K(1-M/100)}$$

Or $$K_1=\frac{IK(1-M/100)}{I+(KM/100)}$$

Or $$1-\frac{N}{100}=\frac{I(1-M/100)}{I+(KM/100)}$$

Or $$\frac{N}{100}=\frac{\left(\frac{KM}{100}\right)+(IM/100)}{I+(KM/100)}$$

Or $$N=\frac{(K+I)M}{I+(KM/100)}$$

With the reference set of data as before, values of N are shown in Table 8.12 considering relevant parameters:

$$K = 100$$
$$I = 40$$

Values of N are higher than that of M but are less sensitive towards larger values of M.

Table 8.12: Variation of N (shortage cost) corresponding to M (decreased C).

S. No.	M	$N=\frac{(K+I)M}{I+(KM/100)}$
1	2	6.67
2	4	12.73
3	6	18.26
4	8	23.33
5	10	28.00

In case of a setup cost increase, a potential response for similar total cost can be the shortage cost reduction.

$$C_1 = C\left(1+\frac{M}{100}\right)$$

$$K_1 = K\left(1-\frac{N}{100}\right)$$

Now:

$$\sqrt{\frac{2DCKI(1-D/P)}{(K+I)}} = \sqrt{\frac{2DC_1K_1I(1-D/P)}{(K_1+I)}}$$

Or $$\frac{CK}{(K+I)} = \frac{C_1K_1}{(K_1+I)}$$

Or $$\frac{1}{(K+I)} = \frac{(1+M/100)(1-N/100)}{I+K(1-N/100)}$$

Or $$\frac{I+K(1-N/100)}{(1-N/100)} = (1+M/100)(K+I)$$

Or $$\frac{I}{(1-N/100)} = (1+M/100)(K+I)-K$$

Or $$1-\frac{N}{100} = \frac{I}{(1+M/100)(K+I)-K}$$

Or $$\frac{N}{100} = \frac{(1+M/100)(K+I)-K-I}{(1+M/100)(K+I)-K}$$

Or $$\frac{N}{100} = \frac{(K+I)[1+(M/100)-1]}{(KM/100)+I(1+M/100)}$$

Or $$\frac{N}{100} = \frac{(K+I)(M/100)}{I+(KM/100)+(IM/100)}$$

Or $$N = \frac{M(K+I)}{I+(K+1)(M/100)}$$

With the reference set of data as before, values of N are shown in Table 8.13 considering relevant parameters:

$$K = 100$$

$$I = 40$$

Values of N are higher than that of M but are less sensitive towards larger values of M.

While comparing with the previous table, these values are lower. This is because:

$$I+(K+I)(M/100) > I+(KM/100)$$

Table 8.13: Variation of N (shortage cost) corresponding to M (increased C).

S. No.	M	$N = \dfrac{M(K+I)}{I+(K+I)(M/100)}$
1	2	6.54
2	4	12.28
3	6	17.36
4	8	21.87
5	10	25.93

8.5.3 Holding cost

With the holding cost increase, shortage cost reduction can be a response for both objectives, i.e., similar batch size and similar total cost.

$$I_1 = I\left(1+\frac{M}{100}\right)$$

$$K_1 = K\left(1-\frac{N}{100}\right)$$

For similar batch size:

$$\sqrt{\frac{2DC(K+I)}{KI(1-D/P)}} = \sqrt{\frac{2DC(K_1+I_1)}{K_1I_1(1-D/P)}}$$

Or $$\frac{(K+I)}{KI} = \frac{(K_1+I_1)}{K_1I_1} \tag{8.3}$$

Or $$\frac{1}{I_1}+\frac{1}{K_1} = \frac{(K+I)}{KI}$$

Or $$\frac{1}{K_1} = \frac{(K+I)}{KI} - \frac{1}{I(1+M/100)}$$

Or $$\frac{1}{K(1-N/100)} = \frac{(K+I)(1+M/100)-K}{KI(1+M/100)}$$

Or $$\frac{1}{(1-N/100)} = \frac{K(1+M/100)+I(1+M/100)-K}{I(1+M/100)}$$

Or $$1-\frac{N}{100} = \frac{I(1+M/100)}{(KM/100)+I+(IM/100)}$$

Or $$\frac{N}{100} = \frac{(KM/100)+I+(IM/100)-I-(IM/100)}{(KM/100)+I+(IM/100)}$$

Or $$\frac{N}{100} = \frac{(KM/100)}{I+(K+I)(M/100)}$$

Or $$N = \frac{KM}{I+(K+I)(M/100)}$$

With the reference set of data as follows:

P	*D*	*C*	*I*	*K*	*Q*	*E*	*J*
960	600	45	40	100	71.0	760.64	7.61

And considering the relevant data, the values of *N* are shown in Table 8.14. These are much higher than that of *M* but are less sensitive towards higher values of *M*.

Table 8.14: Variation of N (shortage cost) corresponding to *M* (increased *I*).

S. No.	*M*	$N = \frac{KM}{I+(K+I)(M/100)}$
1	2	4.67
2	4	8.77
3	6	12.40
4	8	15.62
5	10	18.52

Now, for similar total cost:

$$\sqrt{\frac{2DCKI(1-D/P)}{(K+I)}} = \sqrt{\frac{2DCK_1I_1(1-D/P)}{(K_1+I_1)}}$$

Or $$\frac{KI}{(K+1)} = \frac{K_1I_1}{(K_1+I_1)}$$

As the above expression is similar to (8.3), similar results can be obtained.

8.5.4 Production rate

In case of production rate increase, shortage cost reduction can be a response for the objectives, i.e., similar batch size/total cost.

$$P_1 = P\left(1+\frac{M}{100}\right)$$

$$K_1 = K\left(1-\frac{N}{100}\right)$$

For similar batch size:

$$\sqrt{\frac{2DC(K+I)}{KI(1-D/P)}} = \sqrt{\frac{2DC(K_1+I)}{K_1 I(1-D/P_1)}}$$

Or $$\frac{(K+I)}{K(1-D/P)} = \frac{(K_1+I)}{K_1(1-D/P_1)} \qquad (8.4)$$

Or $K_1(1-D/P_1)(K+I) = K(K_1+I)(1-D/P)$

Or $(1-N/100(1-D/P_1(K+I) = K_1(1-D/P) + I(1-D/P)$

Or $(1-N/100)(1-D/P_1)(K+I) - K(1-N/100)(1-D/P) = I(1-D/P)$

Or $$1-\frac{N}{100} = \frac{I(1-D/P)}{(1-D/P_1)(K+I) - K(1-D/P)}$$

Or $$\frac{N}{100} = \frac{(1-D/P_1)(K+I) - K(1-D/P) - I(1-D/P)}{(1-D/P_1)(K+I) - K(1-D/P)}$$

Or $$\frac{N}{100} = \frac{(K+I)[1-(D/P_1)-1+(D/P)]}{K[(1-D/P_1)-1+(D/P)] + I(1-D/P_1)}$$

Or $$\frac{N}{100} = \frac{(K+I)[(D/P)-(D/P_1)]}{K[(D/P)-(D/P_1)] + I(1-D/P_1)}$$

Or $$\frac{N}{100} = \frac{(K+I)[(D/(P_1/P)-D]}{K[D(P_1/P)-D] + I(P_1-D]}$$

Or $$\frac{N}{100} = \frac{(K+I)\left[D\left(1+\frac{M}{100}\right)-D\right]}{K[D(1+M/100)-D] + I[P(1+M/100)-D]}$$

Or $$\frac{N}{100} = \frac{(K+I)(DM/100)}{K(DM/100) + I[P(1+M/100)-D]}$$

Or $$N = \frac{(K+I)DM}{(KDM/100) + I[P(1+M/100)-D]}$$

With the reference set of data as follows:

P	*D*	*C*	*I*	*K*	*Q*	*E*	*J*
960	600	45	40	100	71.0	760.64	7.61

The values of *N* are shown in Table 8.15. These are much higher than that of *M* but are less sensitive towards higher values of *M*.

Table 8.15: Variation of *N* (shortage cost) corresponding to *M* (increased *P*).

S. No.	*M*	$N = \frac{(K+I)DM}{(KDM/100) + I[P(1+M/100) - D]}$
1	2	10.26
2	4	18.32
3	6	24.82
4	8	30.17
5	10	34.65

Now, for similar total cost:

$$\sqrt{\frac{2DCKI(1-D/P)}{(K+I)}} = \sqrt{\frac{2DCK_1I(1-D/P_1)}{(K_1+I)}}$$

Or $$\frac{K(I-D/P)}{(K+I)} = \frac{K_1(I-D/P_1)}{(K_1+I)}$$

As the above expression is similar to (8.4), similar results can be obtained.

Index

L

O

P

Q

R

S

T

U

V